A FIRST LAB IN CIRCUITS AND ELECTRONICS

YANNIS TSIVIDIS
Columbia University

John Wiley & Sons, Inc.

| New York | Chichester | Weinheim | Brisbane | Singapore | Toronto |

ACQUISITIONS EDITOR	Bill Zobrist
MARKETING MANAGER	Katherine Hepburn
PRODUCTION EDITOR	Ken Santor
TEXT DESIGNER	Lee Goldstein
COVER DESIGNER	Carol Grobe
ILLUSTRATION COORDINATOR	Sandra Rigby

This book is printed on acid-free paper.

The paper in this book was manufactured by a mill whose forest management programs include sustained yield harvesting of its timberlands. Sustained yield harvesting principles ensure that the numbers of trees cut each year does not exceed the amount of new growth.

The procedures in this text are intended for use only by students with appropriate faculty supervision. In the checking and editing of these procedures, every effort has been made to identify potentially hazardous steps and to eliminate as much as possible the handling of potentially dangerous materials, and safety precautions have been inserted where appropriate. If performed with the materials and equipment specified, in careful accordance with the instructions and methods in the text, the author and publisher believe the procedures to be very useful tools. However, these procedures must be conducted at one's own risk. The author and publisher do not warrant or guarantee the safety of individuals using these procedures and specifically disclaim any and all liability resulting directly or indirectly from the use or application of any information contained in this book.

Library of Congress Cataloging-in-Publication Data:
Tsividis, Yannis.
   A first lab in circuits and electronics / Yannis Tsividis.
     p. cm.
   ISBN 0-471-38695-2 (pbk. : alk. paper)
   1. Electronic circuits. 2. Electronics. I. Title.
TK7867 .T75 2001
621.3815--dc21   2001017839

10 9 8 7 6 5 4 3 2 1

Preface

This is a book for a lab course meant to accompany, or follow, any first course in circuit analysis and/or electronics. It has been written for today's students, who, as is by now widely recognized, are very different from students of years past, when most of the traditional teaching approaches were developed. Among other things, today's students are simply bored when required to go through a long series of procedures, the only purpose of which is to teach measurement techniques and to verify the theory. They want to see the theories they learned applied to something useful, and they want this right away. Telling them that they will see the application next year, or even next semester, is not good enough. And if students become disaffected in the first lab, a golden opportunity to excite them about electrical engineering may have been lost forever.

This lab manual has the following objectives:

1. As is the case with any lab manual, to support, verify, and supplement the theory; to show the relations and differences between theory and practice; and to teach measurement techniques.
2. To convince students that what they are taught in their lecture classes is real and useful, and to get them involved in several applications they can relate to. Thus, circuits and electronics are combined in the same lab.
3. To help make students tinkerers (at least for the duration of the lab, and some, hopefully, for a lifetime).
4. To make them used to asking "what if" questions and to acting on their own to discover new things.
5. To motivate their further study. The idea is to explore several concepts in a simple way, which can serve the dual purpose of applications and motivation. For example, the experiment on modulation can whet their appetite for a communications course in subsequent years.

Audience

This book is intended for sophomore or junior electrical and computer engineering students who are taking their first lab, either concurrently with their first circuit analysis class or following that class (or even following a term of electronic circuits). It is also intended for first-year students in electrical and computer engineering at institutions that have started a first-year course in circuits and electronics (an increasing trend). Finally, it is appropriate for nonmajors, such as students in other branches of engineering or in physics, for which electronics is a required course or elective and for whom a working knowledge of circuits and electronics is desirable.

Required Student Background

The lab is meant to run concurrently with, or following, any introductory electrical engineering course. Most electronic circuits used here to make the lab interesting and stimulating for beginners are covered in conventional circuit analysis courses (e.g., op amp circuits). A few simple circuits, such as rectifiers, which may not be covered in a circuit analysis course, are introduced in a self-contained manner. The electronics experiments can extend the lecture material and can serve as excellent motivation for a subsequent course in electronic circuits. Alternatively, the lab may be run concurrently with a course in electronic circuits, in which case the students may not need the introductory background provided or may use it as a concise review. No background in frequency-domain analysis techniques is required, so the lab can be run concurrently with any first circuits course if desired. Nevertheless, the subject of frequency response is adequately covered and applied.

What if circuits is not the first EE course? There is considerable discussion by electrical and computer engineering educators about whether the traditional introduction of students to the field through circuits makes sense today. Some schools are trying other approaches for the first course, such as DSP or "light" control systems. This issue is largely irrelevant as far as the use of this book is concerned: the book is intended for use in a first lab for circuits in electronics, whenever its time comes in a given curriculum. In some schools this will not be the first lab; for example, a computer lab might come before it.

Approach

Although in this book I have adopted the current trend of tightly coupling to applications, I have maintained the classical approach of keeping the experiments largely independent (as opposed to making them being part of a larger, kit construction project). In this way, flexibility has been maintained in designing the experiments to reinforce certain important concepts and to give the instructor considerable freedom in choosing which experiments to cover. Also, in this way applications can be introduced early and students can see results right away.

In developing this book, I have experimented a lot with the level of freedom appropriate for the first lab. On the one hand, I have found that complete freedom is not appropriate, as many students do not know how to begin and become stuck very often. In addition, in the course of a "free" lab, it is possible that the students will not run across some important concepts, which they normally should be taught; so some guidance is in order. On the other hand, a completely regimented approach stifles creativity and does not ensure learning; it is entirely possible for a student to blindly follow instructions, do all required parts, and leave the lab without having really understood much. Hence I have opted for a compromise approach, which works best for the large majority of students. There are steps to be followed in each experiment, but many contain questions or suggestions for extending the results, which require the student to act rather than passively follow. I have spent a lot of time in finding ways to keep students alert and creative in the course of the experiments, often by selectively withholding parts of the story and requiring the students to search for these parts themselves. In other words, both in the choice of experiments and in the format and degrees of freedom within each experiment, I have found that what works best is a mixture of the classical and modern approaches.

The experiments are written to help the student develop intuition and to relate, as much as possible, what is learned or measured to what is perceived through one's senses. For example, in Experiment 3, which deals with time-varying signals, the students are introduced to the function generator and the oscilloscope. Through an amplifier and loudspeaker, they hear the waveforms they observe on the scope's screen; and through a microphone, they observe the waveform of their voice, whistling, or clapping. They are even asked to remove the speaker's panel, touch the paper cone of the speaker very lightly, feel its vibration for various frequencies and amplitudes, and observe how a small particle bounces when placed on the vibrating cone. This may sound overdone, but I know, from my early start as a hobbyist-experimenter, that such sensory experiences stay in memory and help make things click. They provide the confirmation that what is done in the lab is real. This removes psychological blocks, increases intuition, and motivates further study.

Incorporating Applications

I have spent considerable time in identifying suitable applications for illustrating the principles and making them exciting, and I have woven these applications into the experiments. The circuits discussed are connected to applications as soon and as often as possible. Thus, students already see sensors (a thermistor and a photoresistor) in Experiment 2 and use them to design simple temperature- and light-sensitive circuits; they see and use more sensors (microphones) and an actuator (loudspeaker) in Experiment 3; and so on. They do not just measure the gain of an op amp/resistor amplifier, but they use this circuit to amplify their own voice signal and listen to it. They apply diodes to a simple demodulator and LC circuits to receiver selectivity, and they are introduced to wireless communications by building and testing a simple radio receiver (which puts together many of the concepts they have learned up to that point). They do not just measure the frequency response of low- and high-pass filters but also apply the latter to audio tone control, using them to process music from their favorite CD and listen to the result.

Choice of Experiments

There are more experiments in this book—sixteen of them—than can be comfortably covered in one semester. The experiments are designed so that they can be completed within 3 hours each, although some students can finish some of them in about 2 hours. Other durations and adaptations of the experiments to different student backgrounds are possible. Certain parts of each experiment can be omitted if desired (although it would be a pity to omit the application parts, which are what the students are especially looking forward to). Also, parts of different experiments can be combined to form a new experiment. Suggestions are given in the *Instructor's Manual*. I will also be happy to discuss with individual instructors their teaching needs and offer suggestions for putting together a lab course based on this book.

Design Projects

The type of "what if" questions asked throughout this manual encourage the student to experiment and build circuits of his or her own. The book makes possible the introduction of design projects at several points, if the instructor decides that there is room for them. Such projects are appreciated by the students and, if placed between experiments, can be useful as "fillers" for delaying some experiments until the lecture class on the corresponding theory has caught up with them. Several project possibilities, which I have tried over the years, are described in the *Instructor's Manual*.

Lab Equipment Required

The book is designed for a lab that uses equipment as simple (and thus inexpensive) as possible. The basic instruments are two dual power supplies, two digital multimeters, an oscilloscope, two function generators, a inexpensive CD player, a small power amplifier, a microphone, a loudspeaker with enclosure, and assorted parts and cables (an equipment and parts list is given in Appendix G). All this equipment is widely available. For breadboarding, the ubiquitous proto boards can be used, although for the first few experiments at Columbia we prefer Plexiglas boards with plug-in leads so that beginners can clearly see all the connections. Simple instructions for making these boards, if desired, are given in the *Instructor's Manual.*

Instructor's Manual and Web Site

The *Instructor's Manual* will be made available by the publisher to those instructors who adopt the book. The manual contains further discussions of lab equipment, including how to select it; suggestions for running the various experiments; design projects; tips for selecting appropriate teaching assistants for the lab; and other useful information.

A web site, **www.wiley.com/college/tsividis,** will contain extra information, updates, and suggestions.

Class Testing of the Material in This Book

The material in this book has been extensively class-tested over the course of four years with hundreds of students, who filled in detailed questionnaires at the end of each lab session. Every experiment has been revised at least three times. In the last two years, the lab has been run with new teaching assistants and almost no supervision by me; it has run exceedingly smoothly, confirming that by now the experiments in the book are well tuned and foolproof.

At Columbia, the book is used in a lab offered in conjunction with an introductory circuits and electronics lecture course. Students considering electrical and computer engineering are invited to try the course, to see if the discipline is for them. Within the first three years of the course, the yearly number of students who choose electrical or computer engineering as a major had doubled. This is attributed in large part to this first lab. We also found that, on the average, students who have taken the lab perform better in subsequent courses, even those that teach theory. This is not surprising, as one of the main purposes of the lab is to motivate further study.

Acknowledgments

I would like to thank Ron Rohrer, who in an illuminating speech at ISCAS several years ago, discussed the differences in today's students and lit the spark that led to this book. I would also like to thank the colleagues who reviewed this book during the course of its development, and provided useful suggestions and comments, especially John Choma, Jr., University of Southern California; Aniruddha Mitra, Western Nevada Community College; Wai Tung Ng, University of Toronto; Rebecca Richards-Kortum, University of Texas, Austin; Jorge J. Santiago-Avilés, University of Pennsylvania; Jeffrey S. Schowalter, University of Wisconsin, Madison; Ken Shepard, Columbia University; Timothy N. Trick, University of Illinois at Urbana-Champaign; Michael J. Werter, University of California, Los Angeles; and Charles Zukowski, Columbia University. Thanks are due to the Deans of the Fu Foundation School of Engineering and Applied Science at Columbia, Zvi Galil and Morton Friedman, for their support for this project, and to John Kazana, lab manager, for his help in setting up the lab and for his support. Thanks are also due to Nagendra Krishnapura, who as a teaching assistant tried out the first versions of the experiments, and to the many other teaching assistants over the years for conducting the lab sessions and for their patience as the material was undergoing class testing and revisions. Finally, thanks are due to the many students who, with their enthusiasm and detailed comments over four years, have helped to improve the experiments and have made offering the first lab a truly rewarding experience for me.

YANNIS TSIVIDIS

Contents

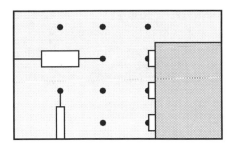

GOOD LAB PRACTICES AND OTHER USEFUL HINTS

You need to know several things before even starting the first experiment in this book. Good lab practices are the key to safe and successful experiments. This chapter contains many important suggestions for you. You may not be able to understand everything at first, as some of the suggestions are given in a context you will encounter later on. Nevertheless, you should read the chapter carefully in its entirety now, so that when the situation arises you will be aware of this material. You should then come back to this chapter and reread as appropriate.

Safety

It is imperative to minimize the dangers of receiving an electric shock by following certain safety procedures. The effects of electric shock are determined by the value of the current that passes through the body, the frequency, the path followed by the current, the time the current persists, and so on. The effects of an electric current can vary from a startling reaction (with unpredictable results) to involuntary muscle contraction (resulting in the "can't let go" effect) to pain, burns, fainting, heart failure, respiratory paralysis, and death.

The amount of current that passes through the body is determined by the voltage applied to it and by the resistance through which the current flows. The resistance can become especially small if there are cuts, if the skin is wet or moist, and if the contact area is large (e.g., through a metallic object such as a watch or a bracelet). Furthermore, the resistance decreases once a current begins to pass. When the body resistance is small, even moderate amounts of voltage can cause a harmful amount of current. It is therefore wise to consider any voltage as dangerous and to take precautions to limit the chances of receiving an electric shock, as well as to be familiar with what to do in case one of your colleagues receives a shock.

- Before starting to work in the lab, familiarize yourself with the location of the circuit breakers and know where to call for help and what to do in case one of your colleagues is injured. Consult appropriate posters or leaflets. If in doubt, ask your instructor.

- Never work in the lab alone.

- Use equipment with three-wire power cords and with a properly grounded case (see the following chapter on ground connections).

- Inspect all cords, plugs, and equipment for possible damage, and notify your instructor if you see any such damage. Also notify your instructor if you see any other sign of trouble, such as loose wall sockets or sparks, or if you receive an electric shock, however small.

- Be careful when inserting or removing a plug. Do not remove plugs by pulling on the cord.

- While making connections, keep the power off.

- Do not touch bare wires and parts. If you have to do so, turn off all power first and unplug the equipment. Even then, be aware that capacitors can store electric charges and can give you an electric shock, especially if their capacitance is large and they are charged to a large voltage.

- Do not work when your skin is wet. Wear shoes while working, and be sure they are dry.

- Do not wear metallic objects such as bracelets, necklaces, rings, or chains while working.

- Do not lean against metal surfaces, such as the case of a piece of equipment, pipes, or the frame of your lab bench.

- Do not leave lose metal parts, including wires, on your bench.

- Do not place drinks or food on your bench.

- If somebody else in the lab receives an electric shock, immediately turn off the power and/or remove the victim from the source of electricity *without* coming into electrical contact yourself (e.g., use a dry piece of wood, a dry piece of cloth, or a nonmetallic belt). Follow appropriate procedures for calling for help, providing artificial respiration and/or cardiopulmonary resuscitation, and otherwise providing aid until medical help arrives.

Possibilities for hazards other than electric shock also exist in an electronics laboratory. Do not exceed the voltage and/or power rating of electronic components. Be very careful to observe the polarity of electrolytic capacitors, which can explode if connected in the wrong way. Be aware of the fact that electronic components can overheat. Be careful with sharp objects, such as wire ends, component terminals, and integrated circuit pins. Be careful with soldering irons; they can cause burns or fires. Long hair and loose clothing can cause hazards when tangled with circuit boards, equipment, soldering irons, or machinery with moving parts.

Additional considerations and rules may apply in your situation and environment. If in doubt, consult your instructor.

Wiring Your Circuit   There are several ways in which you can build an experimental circuit (as opposed to a permanent one, meant for repeated use). A common way to conveniently assemble a circuit without having to solder is to use a plastic "proto" board. This section has been written for the use of such boards, but many of the wiring suggestions given here apply to other types of boards as well.

A plastic proto board has sets of holes into which wires can be inserted. Inside each hole, invisible to you, is a metallic socket appropriate for snugly receiving a wire pushed into the hole. Sets of sockets, arranged in rows and columns, are connected

Fig. 1

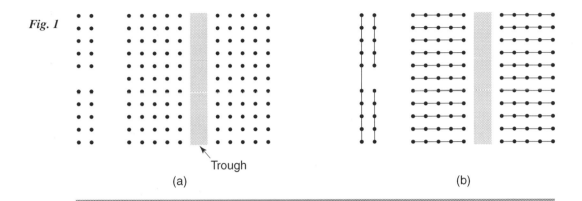

Trough

(a)   (b)

together internally. An example of part of a proto board is shown in Fig. 1. The holes as they appear externally are shown in (a); the internal connections under the surface are revealed in (b). Each set of holes connected together as shown can form a single node in a circuit. Most sets contain five or six holes and are used for internal nodes in the circuit. Other sets, such as the ones shown vertically in (b), are used for power supply and ground connections. On some boards, some of these sets may be broken into separate parts, as indicated in the second vertical line in (b). There are several variations of proto boards, more or less based on this general scheme. Your instructor can explain the internal connections of the particular board you will be using (or you can find them out yourself by using an ohmmeter as a continuity tester, as explained in Experiment 2; you can attach small pieces of wires to the ohmmeter's probes and insert those into the holes). Proto boards are appropriate for low-frequency work (up to a few MHz). At higher frequencies, the large capacitance between adjacent rows of connector holes can interfere with proper operation.

To make connections to a proto board, use pieces of insulated solid (not stranded) wire of an appropriate diameter (usually #22 to #24 AWG), the insulation of which has been removed on both ends, exposing about 12 mm (or about 0.5 inches) of bare wire. Several different lengths of such wires may be appropriate for a given circuit. Sets of connecting wires are commercially available in various lengths and colors. To insert a connecting wire, make sure its bare ends are straight and push each end vertically all the way into a hole. A pair of long-nosed pliers can help in this task. Electronic components such as resistors, capacitors, or integrated circuits can also be plugged directly into a proto board. In the case of resistors, common ½ W ones have lead diameters that are well suited for this purpose. Resistors with a larger power rating may have leads that are too thick, which can damage the connections in the board.

It is very important to get into the habit of wiring a circuit neatly right from the beginning. A neatly built circuit is less likely to contain mistakes, it is easier to debug, and it is easier for a colleague or for your instructor to understand. Figure 2(a) shows a neatly wired circuit, whereas Figure 2(b) shows a messy one. In principle, both wirings implement the same circuit; but if your wiring habits are like those in Figure 2(b) you will soon run into trouble. Several hints to help you form good wiring habits follow. You can interpret several of them by referring to Figures 2(a) and 2(b).

■ Keep the power off while you are wiring your circuit, as well as while you are changing anything in it.

■ Always start with a carefully drawn schematic of the circuit you want to build, properly labeled with component types or values and pin numbers for integrated circuits. Do not try to do things directly from memory.

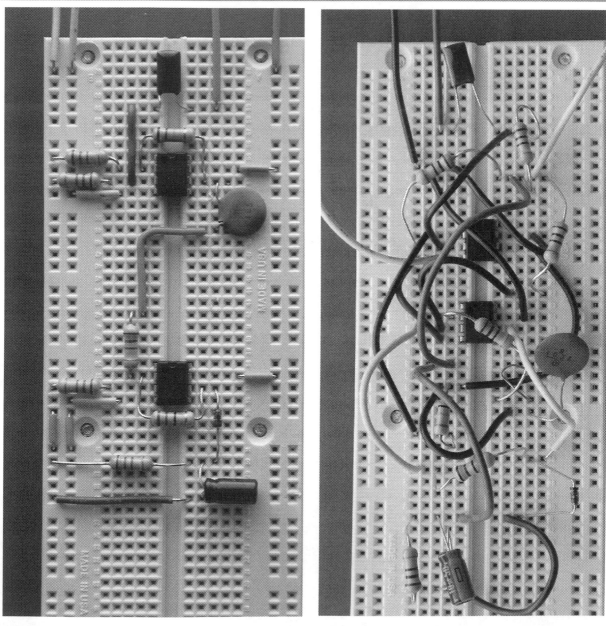

Fig. 2   *(a)*   *(b)*

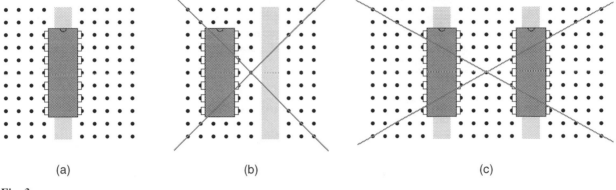

(a)   (b)   (c)

Fig. 3

- Mark all completed connections on your diagram as you go, for example, by using a color marker.

- Use connections as short as possible. Long wires contribute to messiness and can cause interference through undesirable capacitive, inductive, or electromagnetic interaction with other parts of the circuit. (This suggestion does not imply that you need to cut each wire exactly to size; for this lab, it will be sufficient to choose the right length among several pre-cut wires commonly found in a wire kit.)

- Keep wires down, close to the board's surface.

- As a rule, if you can connect components such as resistors or capacitors directly (without extra wires connected to them), do so.

- Plug in chips so that they straddle the troughs on the proto board. In this way, each pin is connected to a different hole set, as shown in Fig. 3(a). Fig. 3(b) shows a common mistake (for a six-hole set proto board). Why is this practice wrong? You can answer this question by considering how the holes are connected together internally [Fig. 1(b)]. You can also see why the practice shown in Fig. 3(c) for a two-trough board is most likely inappropriate when working with more than one chip.

- Do not pass wires over components (or over other wires, if you can avoid doing so without significantly lengthening a connection). This makes the circuit difficult to figure out, and it makes it difficult to remove a component if you have to replace it.

- Use the longer strings of holes as power and ground "buses." For example, use one such string for the connections to the power supply pins of your chips and other components, and wire that string to your power supply.

- Be sure that bare wires or component terminals are clear of each other so that they cannot become accidentally shorted together if something is moved.

- Do not use more wires than you have to. The more connections, the more likely it is that something can go wrong (e.g., a wiring error can occur, or a connection in the proto board can be loose).

- Use color coding for your wire connections (e.g., red for all wires connected to the positive power supply and black for all wires connected to "ground"). This makes it easier to inspect and debug the circuit.

■ If convenient, locate components associated with a circuit with a well-defined function, close to each other in one group. For example, if your circuit includes a preamplifier, all components that are part of the preamplifier should be physically located close to each other, in an easily identifiable block. This can help later during the debugging process (see below).

■ Before plugging ICs (integrated circuits) into a proto board, be sure that their pins are straight. The tips of the two rows of pins on the ICs you will be using in this lab should be 0.3″ apart. Sometimes you will find some ICs with their pins bent; in that case, straighten them before plugging in the ICs. A pair of long-nosed pliers can make this task easy, as then each row of pins can be straightened at one time.

■ Be aware of the fact that ICs, especially those made with MOS technology, can be instantly damaged by static electricity, such as that accumulated by your body. You should make sure you are "discharged" before handling them, by momentarily touching the metal case of a properly grounded instrument.

■ Be sure to use resistors with a sufficiently high power rating (½ W resistors are fine in almost all circuits in this lab) and capacitors with a sufficiently high voltage rating. Observe the polarity on electrolytic capacitors—they can explode if connected in the wrong way.

■ Be sure capacitors are discharged before plugging them into a circuit. If in doubt, short their leads (in critical cases you may have to do so at least twice, as a capacitor can have residual charges even after you have discharged it once).

■ Unplug ICs very carefully, to avoid bending the pins. You may want to try an IC extractor if one is available in your lab.

■ For points that must be connected to external instruments such as power supplies and function generators, connect a wire from an appropriate hole on your proto board to a sturdy post (such as a banana or coaxial socket, often available on the larger test board on which a proto board may be mounted). Then, use an appropriate cable to connect that post to the instrument. This helps make your connections mechanically stable.

■ For the inputs of measuring instruments, such as voltmeters and oscilloscopes, you should normally follow the same practice as above. An exception to this rule occurs when you want to probe several different points in your circuit by moving the instrument probe from point to point or when the connection to the instrument probe must be as short as possible. If you have to connect an instrument probe to a hole on your proto board, you can do so through a short piece of wire; be sure the connection of this wire to the probe is not loose and does not touch any other connections.

■ The coaxial connector of oscilloscope probes (Experiment 3) should only be connected to the input of the oscilloscope. It should never be connected to the function generator, to other instruments, or to your test board.

■ Be especially careful of ground connections. Read carefully the chapter on ground connections.

■ When finished wiring a circuit, inspect all connections to make sure you have made no mistakes. Only when you are happy with the result should you turn on the circuit.

■ If you have connected a signal source (such as the output of a function generator), do not turn it on until after you have turned on the power supplies, if any are used in your circuit. Some ICs can be damaged if you do otherwise. If you later want to turn off the power, turn off the signal source first. In short, never have a signal connected to a circuit with power supplies unless those supplies are on.

Debugging

It is very likely, despite the care with which you may have wired your circuit, that it will not work when you first turn it on, perhaps because of a defective component or a wrongly designed circuit; often, however, this is due to a missing or wrongly connected wire, a loose connection, bent IC pins that have not been properly inserted into holes on the board, an unintended short between two bare wires or terminals, and so on. You will then need to debug, or troubleshoot, your circuit. Trying to do so for a large circuit can be a very difficult task (made easier with experience). Following are some tips to help you debug your circuit.

■ Turn off your circuit and visually inspect it for wrong connections or accidental shorts.

■ Be sure that you know the correct underlying connection pattern of your proto board and that you have not made any mistakes in this respect. Beware of breaks in some power supply or ground buses on some boards [see Fig. 1(b)]. Beware of mistakes such as those shown in Figs. 3(b) and 3(c).

■ If you suspect that a connection is not made as intended, you can use an ohmmeter as a continuity tester (see Experiment 2), but first make sure that all power to the circuit has been turned off.

■ A malfunctioning circuit may have overheated components. Be careful with them. On the other hand, a hot component may be a clue to the problem.

■ If your power supplies have a current meter, observe it. If the current is zero, it may be a sign of a missing connection. If it is excessive, it may be a sign of a short or other problem.

■ In some power supplies you can a set a current limit. You may want to limit the current accordingly, to protect your components in case of circuit errors. On the other hand, if this limit has been set too low (lower than the expected total current for the circuit), it may be the very reason the circuit does not work properly.

■ Use a voltmeter to test whether the power supply voltage reaches an intended point in the circuit, attaching one of its terminals to ground and the other to that point. Test all such points if necessary. A zero voltage may indicate a broken connection or a short to ground. If a voltage is correct at some point, move your probe to the next point where this voltage, or part of it, is supposed to appear, and remeasure. It is usually better to do this type of testing with only the power supplies on and any signal sources off.

■ If the preceeding procedure does not reveal the problem and if your circuit contains a signal source, turn it on and do some further sleuthing with the oscilloscope: Is the signal present at the output of the signal source? If so, is it present at the input of your test board? If not, you may have a bad cable or a short to ground. If there is a signal at the input of the board, is it also present at the next logical point (e.g., at an IC pin to which the board input is supposed to be connected)? The idea is to eliminate possible problems one by one until you hit the points that are causing the trouble.

■ Unless the circuit is very simple, do not try to check its behavior all at once. Rather, do some sleuthing to find out where the problem lies. If the circuit can be broken down (physically or mentally) into two independent parts, check each part separately (e.g., the preamplifier and the power amplifier of the sound system in Experiment 5). To do this, you may need to power the part being tested separately from the rest of the circuit, apply to it an appropriate signal (e.g., obtained from the function generator), and check its output accordingly (e.g., using an oscilloscope). The use of the function generator and oscilloscope is described in Experiment 3.

■ You can carry the above hint one step further: When wiring a large circuit, you can first wire a part of it and test it by itself to make sure it is correct before proceeding to wire the rest. For this approach to work, of course, you need to be sure that the part you are testing is independent of the rest and is supposed to work correctly by itself.

■ If you must try to find out what the problem is by changing things on your test board, change them one at a time, observing the result in each case. If you change more than one thing at once, not only may you be unable to isolate the cause of the problem, but also the second change may undo the result of a first, correct change, and you will have missed your chance to make the circuit work.

■ In some cases problems can be caused by external interference (from a TV station, from lighting or other equipment in the room, etc.). Such interference can be picked up by long wires connected to your circuit (e.g., those carrying power to your setup), which can act as antennas. In such cases, you may be seeing a signal at the output of your circuit, even though no signal is being applied at its input. If you suspect interference problems, you may have to combat them by using bypass circuits, as explained in Appendix D.

■ If all else fails, and only as a last resort, you may have to disassemble your circuit and start from scratch, especially if sloppy wiring in your first version prevents you from identifying the problem. This time, it is hoped, you will be more careful and neat with your wiring.

■ This last solution should not be overused. You should stick with your original circuit and persist in looking for the problem. You can learn a lot not only from properly working circuits but also from knowing what went wrong in bad ones.

Lab Reports   There are many types of lab reports, which vary according to the kind of information that has to be reported, the level of informality, and so on. Your instructor will let you know what type of report he or she expects. Unless your instructor indicates otherwise, it is wise to adopt the following practices.

■ Use a pad of quad graph paper (with squares ¼″ on a side). The squares will make it easier for you to construct tables and plots and to draw schematics.

■ Be sure your report is orderly and neat. Conciseness is appropriate for some styles of reporting, but sloppiness isn't. Even *you* may find it hard to follow your own sloppy report a few weeks after you have written it.

■ Include a careful schematic for each circuit you have built, labeled with component values, types and pin numbers of ICs, and other pertinent information. This is needed for anyone (including your instructor, your colleagues, or even yourself at a later time) to be able to interpret what you have done. In some cases, you may even want to include a board-level layout of your components and connections, corresponding to their physical location on your proto board.

■ When preparing plots, label your axes appropriately, including tics and values along them; the quantity being plotted along each axis; and the corresponding units.

■ Show experimentally obtained points as dots or small circles on plots. Be sure to take a sufficient number of points to appropriately show the behavior you are investigating, especially in regions where rapid variations are observed. Points that are far apart may completely bypass such regions.

■ Pass a best-fit line through the points on your plot. Do not simply join them with straight-line segments.

■ When asked to plot a variable y versus a variable x without further specification, consider what ranges and values are appropriate. When asked to plot current versus voltage for a resistor, for example, it is appropriate to use both positive and negative voltage values unless instructed otherwise.

Preparation

Always study the experiment you are going to do before coming to the lab. Read each instruction carefully, trying to imagine your experimental setup. Try to anticipate the probable results of your measurements, as well as what to watch out for and what problems are likely to arise. Take notes as appropriate. Being fully prepared is necessary for you to be able to finish an experiment in the time allowed and to be able to benefit from it as much as possible. The experiments in this book are based on the assumption that you will have fully prepared yourself before doing each one. You also owe it to your lab partner to be fully prepared, so that you can contribute to the experiment.

Running the Experiment

■ Always read each step in its entirety before acting. Do not stop in the middle of a step and start wiring or measuring, as the rest of the step may contain information relevant to those tasks, which can save you time or trouble.

■ Try to guess the likely result of each step before you perform it. This will enhance your understanding and will prepare you to catch a possible problem, saving time.

■ Do not put off plotting or other tasks until the end of the experiment. Do plots when asked to. This is because some plots reveal information that can be useful in interpreting your results and identifying potential problems before time is wasted on a possibly malfunctioning circuit. Plots can also tell you whether you need to measure more points before going further, while you still have your circuit connected for doing so. Finally, they can give you intuition and information that will help you in the subsequent steps. Again, completing each task when asked to in each experiment can save time and trouble.

■ Measurements are never exact. Keep this in mind, but otherwise try to obtain measurements as accurately as possible. Think about the sources of measurement error in each case, and interpret your results accordingly.

■ Use an appropriate range on measuring instruments to obtain enough significant digits. On the other hand, you should not overdo this. It is not appropriate to give the impression of precision by keeping more digits than makes sense in a given situation. Also, with some instruments (such as some ammeters; see Experiment 1), the range chosen can affect the degree to which the instrument affects the circuit under measurement.

■ Be very careful if you need to obtain a small quantity by subtracting two much larger quantities. For example, $1.344 - 1.336 = 0.008$, but if the two numbers had been measured with three significant digits, you would have gotten $1.34 - 1.34 = 0$.

■ Become very familiar with your test instruments. Do not arbitrarily push buttons until you get something; *know* which buttons to push.

■ If you are doing an experiment with a lab partner, make sure you both contribute equally. If you do not contribute, you are not only being unfair to your partner but also not really learning. Passive observation cannot replace doing. If, instead, you are the type that takes over and would rather do the lab by yourself, you are hurting your partner's learning process. Be sure both of you have a chance to do each type of task. For example, if you wire and your partner takes measurements, the next time around your partner should do the wiring and you should measure.

■ Above all, **resist the temptation to just blindly follow the procedures.** If you just do so, take all the measurements correctly, and write your report, you will have wasted several hours. Observe, think, act, and discover. Many of the "why" questions in this book are meant to just make you think. But do not stop there. Ask yourself, as often as you can, why something is done in a certain way, why it works or doesn't, or what would happen if something were done differently. This is a very important part of your education. Discuss such questions with your lab partner. If you have ideas that you want to try, first make sure they are safe; if in doubt, ask your instructor.

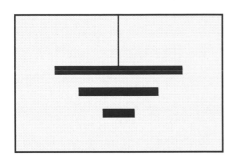

GROUND CONNECTIONS

Introduction

The issue of ground connections is one that will concern you again and again, in this and in other labs. Read this chapter carefully, and try to understand it as much as possible. Not everything in it may make perfect sense in the beginning; some of this material will become clearer as you gain laboratory experience. Nevertheless, it pays to have a preliminary understanding at this point. As you attempt to connect various instruments in future experiments, you may need to return to this chapter for advice.

Producing positive or negative supply voltages with respect to ground

Lab instruments have terminals so that you can connect them to the circuits you are working with. For example, a "floating" power supply has plus (+) and minus (−) terminals. The voltage between them is a well-defined quantity, which you can set at will. However, the voltage between one of these terminals and a third point, such as the instrument's metal case (if it has one) or the case of another instrument, may not be well defined and may depend, in fact, on instrument construction and parasitic effects that are not under your control. Parasitic voltages can interfere with proper operation of the circuits you are working with, or can even damage them. Worse, in some cases they can cause an electric shock. To avoid such situations, the instruments have an additional terminal called the *ground,* often labeled GND or indicated by one of the symbols shown in Fig. 1; the use of this terminal will be explained shortly. The ground terminal may be connected to the internal chassis of the instrument (it is sometimes referred to as the chassis ground); to the instrument's metal case; and if the power cord of the instrument has three wires, to the ground lead of the cord's plug. When you plug in the instrument, this lead comes in contact with the ground terminal of the power outlet on your bench, which is connected to earth potential for safety and other reasons. In fact, other instruments on your bench, on other benches, or even elsewhere in the building may have their grounds connected to that same point, through the third wire of their power cords.

Fig. 1

When you use a power supply with the output floating (i.e., with neither of the output terminals connected to ground), you get the situation shown in Fig. 2. The little circles indicate terminals for making connections to the instrument's output or ground. In the following discussion, V_{XY} will denote the voltage from a point X to a point Y. In Fig. 2, V_{AB} is the power supply's output voltage V, and it is well-defined. However, V_{AG} and V_{BG} are *not* well defined and can cause the problems already mentioned. To avoid this, you should strap one of the two output terminals to the ground terminal, as shown, for

Fig. 2

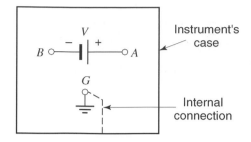

example, in Fig. 3(a). Now *all* voltages are well defined: $V_{AB} = V$, $V_{BG} = 0$, and $V_{AG} = V_{AB} + V_{BG} = V + 0 = V$. If we assume that V is a positive quantity, the connection in Fig. 3(a) develops a positive voltage at terminal A with respect to ground. If you happen to need a negative voltage with respect to ground instead, you would use the connection in Fig. 3(b). Here terminal B has a potential of $-V$ with respect to ground. In some power supplies, ground connections as shown in Fig. 3 are permanent, and you do not have access to them. In other power supplies, it is up to you to make such connections.

Connecting one grounded instrument to another

When more than one instrument or circuit with ground connections are used, one should think carefully. Consider the situation in Fig. 4, where it is attempted to connect the output of one instrument to the input of another. For example, instrument 1 can be a function generator, discussed in Experiment 3. Instrument 2 can represent an oscilloscope, or an oscilloscope probe, also discussed in Experiment 3. At first sight, the connections shown seem to be correct. However, there is a *big* problem. Although not apparent from Fig. 4, the ground terminals are connected not only to the instrument cases *but also to the common ground of the power outlet on the bench* (through the ground pin on the power plug, as explained earlier). Making these connections explicit, we have the situation shown in Fig. 5. It is now clear that the second instru-

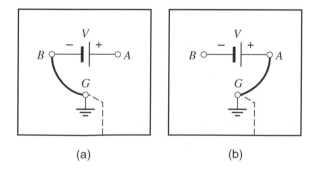

Fig. 3   (a)   (b)

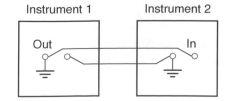

Fig. 4

ment's ground connections short the first instrument's output across *CD* (i.e., they place a short circuit across it; you can trace this short circuit along the path *CIHGED*). Not only will this prevent a voltage from being developed at that output, but also it can damage the instrument. The problem is solved if the connection between the two instruments is modified, so that *instrument ground is connected to instrument ground,* as shown in Fig. 6.

At this point, one may wonder what the connection marked *x* is needed for in Fig. 6, given that the two ground terminals are connected together anyway through the power cables, as shown by the heavy lines. The answer is that there may not always be a ground terminal on the power plug, and even if there is one, it may not be reliable; although ideally all ground terminals on power receptacles should be at the same potential, they sometimes are not. In addition, the long ground wires (*IH* and *GH* in Fig. 6) may act as antennas, picking up interference. To be safe, then, use a short connection such as *x* between the ground terminals of the two instruments.

A final word of caution: Since an instrument's case is in contact with ground connections, you need to be sure that cables and devices do not accidentally come into contact with it. If this happens, malfunction or damage can occur.

These guidelines will be sufficient for the purposes of this lab. Grounding is actually a complicated issue, and you should not expect the simple practice discussed above to be adequate in all situations. As you gain experience, you will obtain a better feel for grounding practices.

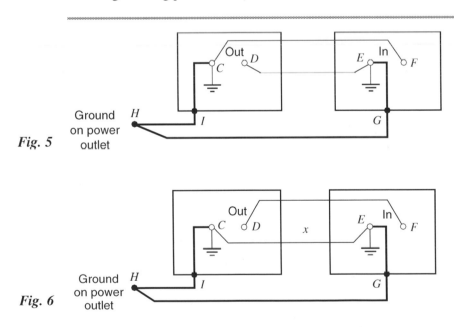

Fig. 5

Fig. 6

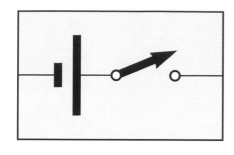

EXPERIMENT 1
MEASURING DC VOLTAGES AND CURRENTS

Objective   This experiment will familiarize you with the measurement of voltage (v), current (i), and v-i characteristics; it will also give you a feel for these quantities and for the important notion of reference direction.

Preparation   Read about the definitions of voltage and current in your text. Pay special attention to reference directions for these quantities. Study the chapters entitled "Good Lab Practices and Other Useful Hints" (paying special attention to the section on *Safety*), and "Ground Connections" in this manual.

INTRODUCTION

The two instruments you will be using in this experiment are the power supply (PS) and the digital multimeter (DMM). The PS will be used to power the circuits you will be building. The DMM is one of the most important instruments in the EE laboratory, and it is one of the simplest to understand and to operate. You will find a PS and a DMM on your bench. Simple user's manuals for both instruments may be made available by your instructor. In this experiment, we will use the DMM to measure DC voltage and current. Study the face of the PS and the DMM, and try to guess the function of each control before proceeding.

The following instructions have been written for a generic DMM. The particular type you will be using may have to be operated somewhat differently. Your instructor will tell you if additional settings, considerations, and precautions apply to the particular DMM type you will be using.

VOLTAGE MEASUREMENTS

1. Set up the DMM as a voltmeter. This will involve *both* pushing the appropriate buttons *and* connecting the DMM's leads (one of which is red, and the other black) at the appropriate DMM terminals. In some DMMs, the terminal to which the red lead should be connected is red itself or is labelled HI, or V; the terminal where the black

Fig. 1

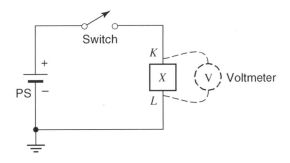

lead should be connected is black itself, or is labeled LO or COM. If you do not see these markings, ask your instructor for help. If there is a DC/AC button on the voltmeter, set it to DC since only DC (constant with respect to time) voltages and currents will be measured in this experiment.

If the DMM is properly set up as a voltmeter, it will behave approximately as an open circuit. Make sure you do *not* set it up as an ammeter (current meter), in which case it will act as a short circuit and, if improperly used, can cause damage.

To measure a voltage between two points, a voltmeter must be connected *across* the two points so that it can sense the potential difference between them. If you want the potential *of* point K *with respect to* point L, you need to connect the *red* terminal of the voltmeter at K, and the *black* terminal at L. You should *not* disconnect anything in the circuit under measurement in order to measure a voltage in it.

2. Connect the − terminal of the PS to the PS's ground terminal. Turn on the power supply (PS). Connect the black DMM lead to the − terminal of the power supply, and the red lead to the + terminal. Measure the voltage of the PS for various settings of its voltage knob. For a reliable measurement, you should set the voltmeter to an appropriate measurement range. The range must be one for which the maximum measurable voltage is higher than the voltage you are trying to measure. Also, the range should be such that a sufficient number of decimal places are shown in the voltmeter display; for our purposes, two decimal places are enough. Experiment with various range settings of the voltmeter, and try to fully understand their purpose and function. What will the problem be if the range on the DMM is too small for the voltage being measured? If it is unnecessarily large? Compare the reading of the DMM to that of the PS's own voltmeter, if there is one. Be sure you interpret the units on the instruments correctly (e.g., mV means 0.001 V).

3. Set the PS voltage at a certain value, and record the reading of the DMM. Then, *without changing the PS and DMM settings*, interchange the connections of the DMM's leads at the PS and record the new reading. How are the two readings related? Why?

4. Set up the circuit shown by the *solid* lines in Fig. 1. Although many devices can in principle be used as element *X* in the figure, for our purposes this element will be a resistor with a value of several kΩ, which will be provided to you. The switch shown

is not part of the PS; you should use a stand-alone switch provided to you. Set the PS voltage at a few volts. Now attach the voltmeter, as shown by the broken line, and measure the voltage of point K with respect to point L, with the switch open and with the switch closed. Now measure the voltage of point L with respect to point K. Relate this reading to the previous one and explain.

CURRENT MEASUREMENTS

5. You will now *prepare* for measuring, in step 6 below, the current through element *X*. First, *disconnect* the DMM from the preceding circuit, leaving the rest of the circuit connected. Configure the DMM as an ammeter (current meter). This will involve both button pushing and selecting the appropriate DMM terminals for connecting the red and black leads. In some DMMs, these terminals may be the same as those used for voltage measurements; in others, the red lead may have to be connected to a separate terminal, which may be labeled A.

> **To measure a current at a given point in a circuit, you need to break the connection at that point and insert the ammeter there. To measure the current in a given reference direction, you need to make the current *enter* the ammeter at the *red* lead, go *through* the ammeter, and *exit* from the *black* lead.**

If a DMM is properly set as an ammeter, it will appear approximately as a short circuit (or "short," for short :-) between the two leads. So, when the ammeter is inserted in a circuit, it does not disturb the circuit; it acts approximately as a wire. In many ammeters, the extent to which they act as shorts depends on the measurement range set on them. In general, the larger the range selected, the more the ammeter acts like a short (i.e., the lower the resistance between its two terminals). In ammeters with only a few display digits, it is better to select the range as large as possible, while still allowing for a sufficient number of significant digits to be displayed. For our present purposes, displaying two or three significant digits will be adequate. When interpreting the ammeter reading below, make sure you take into account unit prefixes, if any are present (e.g., mA means 0.001 A).

6. Consider again the circuit shown in solid lines in Fig. 1. Suppose you need to measure the current in the wire that connects the switch to the upper terminal of element *X*, in the direction from left to right. To do so, *break* that connection, and *insert* the ammeter (A) in series with the element, as shown in Fig. 2 (*think about where the red and black leads should be connected*). Ideally, since the ammeter acts as a short circuit, it is just a piece of wire as far as the circuit is concerned and does not influence the circuit's operation. Nevertheless, the current of element *X* now goes through it, and can be measured. Measure the current i_1 indicated in the figure. Keep the PS and the circuit switch in the off position until you are ready to do this measurement, and then switch them on. When finished with this step, return the switch to the off position.

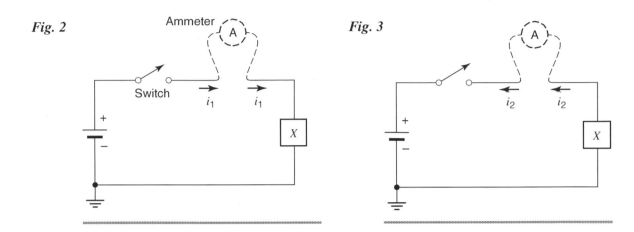

Fig. 2   *Fig. 3*

7. Without changing the settings on the PS and the DMM, you will now prepare to measure the current i_2, indicated in Fig. 3. Notice that the circuit is still the same; only the reference direction for the current has been changed. What should you change in how the ammeter is connected in order to measure i_2? If you are not sure, reread step 5. Now, measure i_2. Compare your reading to that of i_1 obtained above, and explain what you find. Then turn everything off, and disconnect your circuit.

MEASURING V-I CHARACTERISTICS

8. The behavior of resistive circuit elements can be represented by a plot on the voltage-current (v-i) plane. This plot is usually called a *v-i characteristic*. For most resistive devices, there is a unique value of i for each value of v, and vice-versa. To measure i versus v, hook up the circuit of Fig. 4, using a resistive element Y, which will be provided to you (if one is not provided, use a resistor with a value of several kΩ). The arrow through the PS symbol is used to emphasize that the PS voltage is variable. Two DMMs are used, one set up as a voltmeter and the other as an ammeter (alternatively, you can use a single DMM as an ammeter, along with the PS's own voltmeter). Notice that, as before, the voltmeter is connected *across* the element, whereas the ammeter is connected in *series* with the element. Think carefully about where the red and black leads of the DMM(s) should be connected to measure the voltage and current *with the polarity and direction indicated in the figure.*

 Vary the PS voltage and record $v = v_{KL}$ and i, for a voltage range of -1 V to $+1$ V. For negative voltage values, you will need to reverse the $+$ and $-$ terminals on the PS, so that the $-$ terminal of the PS is now connected to the switch, and the $+$ terminal to ground. The PS's own voltmeter will still be showing a positive voltage since it is permanently connected internally to the PS voltage; however, the voltage v_{KL} across element Y will now be negative, and if you have a separate voltmeter connected across this element, you should notice a change of sign in its reading. Plot i vs. v. Use enough measurements to generate a smooth plot.

Fig. 4

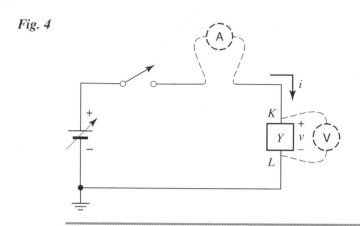

EFFECT OF INSTRUMENT ON CIRCUIT BEING MEASURED

One should be aware that no quantity that can take values in a continuous range can be measured exactly. No measuring instrument is perfect. For example, a real voltmeter is not a perfect open circuit, so a small current can flow through it. Also, a real ammeter is not a perfect short circuit, so a small voltage can develop across it. There are situations in which these small quantities can affect the currents and voltages in the circuit. In such cases, the measuring instruments will be influencing the quantities to be measured. The instrument's indications, then, will have to be properly interpreted, based on knowledge of the instrument's characteristics. We will have a chance to see instances of this in future experiments (e.g., in Experiment 6, we will see how a voltmeter's input resistance can affect a measurement and how we can take this effect into account and correct for it). In most cases in this lab, though, the circuits and element values used are such that the voltmeters and ammeters do not affect the circuits appreciably; they can then be assumed to be ideal, that is, perfect open circuits and perfect short circuits, respectively.

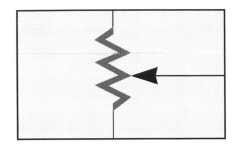

EXPERIMENT 2

SIMPLE DC CIRCUITS; RESISTORS AND RESISTIVE SENSORS

Objective

In this experiment, you will familiarize yourself with series and parallel DC circuits, and you will verify fundamental circuit properties. You will also become familiar with resistors and potentiometers and will learn how to measure resistance. Finally, you will convert nonelectrical parameters (light intensity and temperature) to electrical resistance, using simple sensors.

Preparation

Read about Kirchhoff's voltage and current laws, series and parallel circuits, resistors and combinations of resistors, and voltage dividers in your text.

Suggestion

You will be asked to assemble several circuits below, and to use measuring instruments to measure voltage and current. To avoid confusion, each time follow this practice:

- **First,** assemble the circuit by itself, *without* connecting any measuring instruments (DMMs) to it.

- **Then,** depending on the type of measurement you are asked to perform, *think* how you will connect the appropriate instrument(s) to the circuit; refresh your memory on this point by referring to Experiment 1. In the case of *voltage* measurement, you will simply have to connect a voltmeter *in parallel* with parts of the circuit, *without* breaking any connections; in the case of *current* measurement, you will have to *break* a connection and connect the ammeter *in series* with it.

- **Finally,** once you are sure how to do it, go ahead and connect the instruments.

A SERIES CIRCUIT

1. Construct the series circuit shown in Fig. 1. For this experiment X and Y will be resistive elements provided to you (e.g., two different-valued linear resistors with values of several KΩ; you can learn to read resistor values as explained in Appendix A). Set the

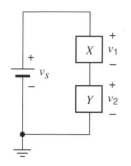

power supply voltage to an arbitrary value. Measure the voltages v_S, v_1, and v_2. (To measure the voltage across an element, you need to connect the voltmeter in parallel with it; you should *not* break any connections to measure the voltages. Think about where the red and black leads of the voltmeter should be connected.) Verify that these voltages satisfy Kirchhoff's voltage law:

$$v_S = v_1 + v_2 \tag{1}$$

Fig. 1

A PARALLEL CIRCUIT

2. Construct the parallel circuit in Fig. 2, using the elements provided for X and Y. Set the power supply voltage at an arbitrary value. Measure the currents i_1, i_2, and i_3. (Do *not* change the PS voltage during these measurements. Think of *which* connection you need to break in each case to insert the ammeter and where its red and black leads must be connected.) Use an appropriate range on the DMM, so that no more than two or three significant digits are displayed. Verify that the currents satisfy Kirchhoff's current law:

$$i_3 = i_1 + i_2 \tag{2}$$

LINEAR RESISTORS AND RESISTANCE MEASUREMENT

3. Plot i versus v for a (linear) resistor specified by your instructor, using the procedure given in Experiment 1, step 8. Use both positive and negative voltage values. Verify that the form of the plot satisfies Ohm's law:

$$i = \frac{v}{R} \tag{3}$$

and find the value of R from the slope of this plot.

4. Turn off the power, and disconnect the circuit you used in the previous step. Configure the DMM as an *ohmmeter*. In this configuration, the instrument passes its own current through a resistor to be measured, measures the voltage across it, and calculates and

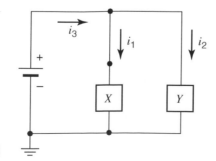

Fig. 2

Fig. 3

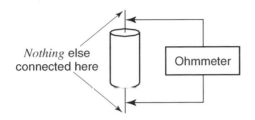

displays the resistance value ($R = v/i$). Measure the resistance of the resistor you used in step 3, using the ohmmeter as shown in Fig. 3. Note: Be sure that *nothing* else is connected to the resistor you are attempting to measure.

Compare the reading of the ohmmeter to the value you found in step 3 above.

5. Using the resistor color code described in Appendix A, read the resistance value. How different (in percentage) is this value from the value measured by the ohmmeter? Is this difference within the specified tolerance for this resistor?

NOTE: In addition to its resistance, an important rating for a resistor is the maximum power it can handle. When using a resistor, one should make sure that the power dissipated on it, $P = vi = v^2/R = i^2R$, is less than that value. For our experiments, ½ W resistors will be more than adequate, and in most cases even ¼ W will be fine.

RESISTIVE SENSORS

6. Sensors are devices that convert nonelectrical physical quantities to electrical ones to make their measurement and handling by electronic circuits possible. In this experiment, you will become acquainted with two types of sensors, which will be provided by your instructor. The first is a *photoresistor*; this is a resistor whose resistance depends on the intensity of light. Verify this property with the ohmmeter, using a photoresistor provided by your instructor. What would be an appropriate way to display its *i-v* characteristics for various light intensities, using a single pair of axes? (Do not provide a quantitative plot; just sketch qualitatively what a proper plot would look like, using the horizontal axis for *v* and the vertical axis for *i*.)

7. Another type of resistive sensor is the *thermistor,* a resistor whose resistance depends on temperature. Repeat step 6, using the thermistor provided by your instructor (any type with a value of several kΩ will do). To change the temperature of the resistor, you can blow concentrated air on it using a straw.

SERIES AND PARALLEL RESISTORS

8. Measure the values of two resistors by using the ohmmeter. Calculate what the equivalent resistance would be if you connected these resistors in series. Now connect them in series, and measure the total resistance. Compare the reading to your calculation.

Fig. 4

9. Repeat step 8 for the two resistors connected in parallel.

CONTINUITY TESTER

10. You will be provided with a "single-pole, double-throw" switch. The symbol for this element is shown in Fig. 4. Depending on the position of the switch, it either connects terminal A to terminal B (as in the figure) or terminal A to terminal C. Use the ohmmeter to identify the unmarked terminals of your switch as terminals A, B, and C in Fig. 4. (Hint: When a connection between two terminals is made, a short is established between them, with virtually zero resistance. When a connection between two terminals is broken, and if no other element is connected between them, an open circuit exists between the terminals, with virtually infinite resistance.) Used as above, the ohmmeter serves as a continuity tester.

VOLTAGE DIVIDER

11. Construct the voltage divider shown in Fig. 5, using two different-valued resistors with values of several kΩ each. Set v_S to any convenient value, measure v, and verify that it satisfies the voltage divider formula:

$$v = v_S \times \frac{R_2}{R_1 + R_2} \qquad (4)$$

12. Design a voltage divider that divides a voltage by a factor of 3 and, when driven by a 10 V supply, dissipates a power no higher than 100 mW. Then connect the circuit and verify its operation.

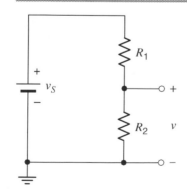

Fig. 5

POTENTIOMETER

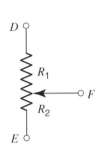

Fig. 6

13. A potentiometer is a voltage divider with a variable division ratio. The symbol for this element is shown in Fig. 6. The total resistance between terminals D and E is constant. The resistance between F and D is R_1, and that between F and E is R_2. The arrow represents a slider, which can slide up and down the length of the potentiometer. (Many practical potentiometers have the resistance laid out over part of a circle, and the slider must be rotated rather than moved up and down.) Thus, the values R_1 and R_2 can be varied, while their total resistance, $R_1 + R_2$, stays fixed.[1] In this way, a variable voltage divider can be implemented, as is evident from eq. 4. Use the ohmmeter to study the potentiometer provided to you. Do *not* connect anything else to the potentiometer in this step. Identify the terminals D, E, and F, and determine the minimum and maximum values of R_1, R_2, and $R_1 + R_2$.

14. Replace the two resistors of Fig. 5 by the potentiometer, as in Fig. 7. Use any convenient value for v_S. Observe the effect of the potentiometer setting on the value of v. What are the minimum and maximum attainable values of v, in comparison to v_S?

CONVERTING PHYSICAL QUANTITIES TO VOLTAGES

15. Propose a circuit that produces a *voltage* that varies with the *light intensity* in the room. Hook the circuit up and try it out. Think of ways to make the circuit sensitive. Demonstrate the circuit to your instructor.

16. Repeat step 15 using *temperature* instead of light intensity.

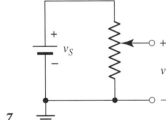

Fig. 7

---

[1]Potentiometers in which R_2 varies in proportion to the slider's displacement from the bottom position in Fig. 6 are called *linear* potentiometers. Another variety is *logarithmic* potentiometers, in which log R_2 is proportional to the above displacement. These are appropriate for use in volume-control circuits because the perception of loudness follows a logarithmic law.

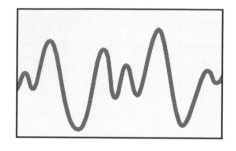

EXPERIMENT 3

GENERATING, OBSERVING, AND HEARING TIME-VARYING SIGNALS

Objective

In this experiment, you will deal with voltages and currents that are not DC, but instead vary as functions of time. You will use an instrument called a *function generator* to produce such signals. The variation of some of these signals will be in the audio range, so you will be able to convert them to sounds and listen to them. You will use a very important measuring instrument, the *oscilloscope*, to observe time-varying quantities (waveforms) and measure their characteristics.

Preparation

Study any material, such as selected pages of manuals or explanation charts that describe the particular function generator and oscilloscope you will be using, that may have been provided to you by your instructor. However, depending on the complexity of the instruments you will be using, it may be possible for you to do this experiment even if no such material has been provided ahead of time. This experiment is designed to guide you so that you can find out what most controls on these instruments do.

Since this experiment has been written with generic instruments in mind, some details (e.g., the precise names of controls) may differ somewhat for the particular instruments in your lab. Your instructor will let you know if different names, settings, considerations, and precautions apply to the instruments you will be using.

THE OSCILLOSCOPE AND THE FUNCTION GENERATOR

1. The oscilloscope (or, simply, "scope") is used to display and measure time-varying signals (waveforms). If you are interested in viewing and measuring a voltage waveform, you can connect a scope to it, just as you connected a voltmeter to a DC voltage in previous experiments. In fact, the first voltage we will display on the scope will be a DC voltage (which is a special case of a waveform).

We will first need to set a number of controls, the function of which will become clear later in this experiment; *for now, do not worry if you do not yet understand what they are.* Depending on the oscilloscope you have, the following settings may be appropriate:

Vertical mode:	channel 1
Channel 1 settings:	position control at midpoint
	sensitivity 1 V/division
	DC input coupling
Horizontal mode:	position control at midpoint
	sweep rate 0.2 s/division
Trigger:	source: channel 1
	DC coupling
	positive slope
	auto mode
	level at midpoint

All continuously variable controls that have a "calibrated" or "CAL" position should be set to that position.

All magnifiers ($\times 10$), if any are present, should be deactivated.

If some of these settings do not make sense for the type of oscilloscope you have, set the controls as specified by your instructor.

2. Turn on the oscilloscope. The display should show a spot, moving from left to right; depending on the type of oscilloscope, this spot may or may not leave a trace behind it. In either case, we will call the result "the trace." *If there is an intensity or brightness control, use it to keep the brightness of the trace low; otherwise, the screen can be damaged in some scopes.* If there is a focus control, use it to make the trace sharp. Use the channel 1 vertical position control to bring the height of the trace to the midpoint of the screen. Adjust the horizontal position control so that the trace starts at the left end of the graticule. Verify that the time it takes for the spot to be swept across the screen is consistent with your setting of 0.2 s/division (a division, or "div," is the distance between two major graticule lines).

3. You will now prepare to connect an external voltage to the input of the scope's channel 1. The outer shield of the scope's input connector is grounded to the metal enclosure of the instrument and to the ground of the electrical installation in the building. A voltage to be measured must be connected between the inner conductor of the input terminal and the outer, grounded conductor. You will be making connections to the scope's input through a $\times 1$ probe (where $\times 1$ means that the probe does not attenuate, or reduce the signal; if your probe has a $\times 1 / \times 10$ selector, set it to $\times 1$). See Fig. 1. The cable of the probe is coaxial, consisting of an inner conductor and an outer shield. The outer shield is connected to the ground and protects the inner conductor from external interference, for example, from a radio station (if two separate wires

Fig. 1

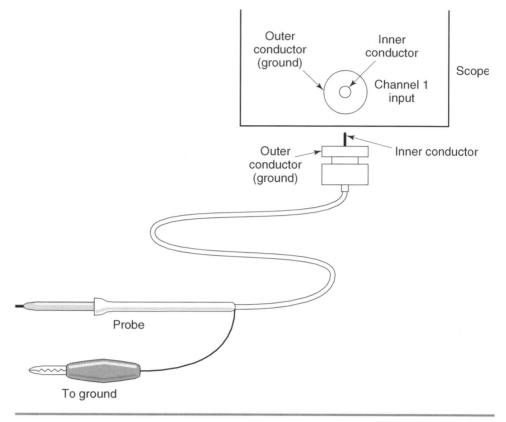

were used instead, they might have acted as antennas, picking up the radio station's signal; this would have interfered with proper measurements).

Connect a $\times 1$ probe to the input of channel 1. Of the two leads of the probe, the flexible one (usually supplied with an alligator clip) is the ground terminal (Fig. 1). **If the voltage source to be measured has one terminal grounded, make sure that the ground of the probe is connected to the ground of the voltage source and not vice versa; otherwise, the source will be shorted and may be damaged.** See the chapter on ground connections in the first part of this book, especially the section entitled "Connecting One Grounded Instrument to Another."

4. To create a voltage source with one terminal grounded and with a positive voltage with respect to ground, first connect the PS's negative terminal to the PS's ground. Now connect the scope's probe to the output of the power supply, *making sure you follow the practice outlined in the previous step.* Turn the supply on, and set its voltage to 2 or 3 volts. Observe its effect on the vertical position of the trace. Vary the setting of the power supply somewhat, and again observe the effect on the vertical position of the trace. Disconnect the probe from the PS. Then, create a voltage source with a negative voltage with respect to ground, and again display it on the scope (*again, be careful with grounds*). What difference do you see?

5. To prepare for measuring the DC voltage with the scope, you need to do two things. *First,* you need to set the *sensitivity* of the scope to an appropriate value. This is like

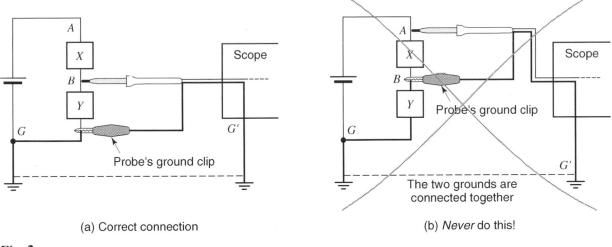

(a) Correct connection   (b) *Never* do this!

Fig. 2

setting the range on a voltmeter. For this measurement, set the sensitivity to 1 V/division. *Second,* you need to establish your zero-voltage reference. Disconnect the probe from the PS, and short together the probe's input terminals to make sure that the scope input voltage is zero (this can be done more conveniently with an appropriate button on the scope's face, as we will see below; for now, though, follow the procedure just described). Use the vertical position control to adjust the height of the trace to the midpoint of the screen. *This position will be your zero-voltage reference; nonzero voltages will be measured with respect to this level.* Now unshort the probe's input terminals.

6.   Again connect the PS to the probe's input. Set the PS voltage at 2 or 3 volts, positive with respect to ground. Measure the PS voltage by observing the vertical displacement of the trace from the zero-reference level:

Voltage (in V) − [displacement (in major divisions)] × [sensitivity (in V/division]

Note that displacement upward is taken as positive, whereas displacement downward is taken as negative. Confirm your measurement by using a voltmeter (the one on the power supply, or the DMM). Change the voltage of the power supply and the vertical sensitivity setting to other values, and repeat the measurement. Be sure it all makes sense to you; at the end of this step, you should feel comfortable with using the scope as a DC voltmeter.

We emphasize here that, since one end of the probe is ground, *it can only be connected to the ground in the circuit to be measured.* Thus *only voltages with respect to ground* can be directly measured by using a probe in the way described above. For example, the connection shown in Fig. 2(a) is appropriate since both the probe's alligator clip and the bottom terminal of device *Y* are connected to ground. However, the connection shown in Fig. 2(b) is **not** allowed since the bottom clip of the probe connects point B to point G'; the latter, being ground, is electrically the same as point G. Thus, device *Y* is shorted out by the two grounds at its top and at its bottom. The operation of the circuit is disturbed, and even damage can be caused by the short.[1]

---

[1]Special techniques can be used to measure "floating" voltages (i.e., between two points, neither of which is ground) with a scope. We will have a chance to see one such technique in Experiment 6.

7. If your probe has a $\times 1 / \times 10$ selector, set it to $\times 10$ and observe the effect on the screen. (If your probe has no such setting, use a separate $\times 10$ probe.) What does this setting do to the overall sensitivity of the probe-scope combination? In this setting, the probe's "loading effect" on the circuit being measured is smaller (i.e., the probe draws less current from it and is less likely to disturb its operation). This setting is used for measuring certain sensitive circuits, and its proper use includes a probe adjustment (tuning), which will not be used here (it is described in Appendix B for later use; you do not need to consult the appendix at this point).

In this experiment, the $\times 10$ setting will not be used further. *Before proceeding, return the probe to the $\times 1$ setting* (or, if you had connected a separate $\times 10$ probe, disconnect that and reconnect the $\times 1$ probe to the scope's input).

8. To prepare for this step, be sure the vertical sensitivity is at 1 V/div, and the sweep time at 0.2 s/div. You will now observe how a time-varying voltage is displayed on the scope. To manually produce such a voltage, grab the voltage control knob on the PS and move it quickly back and forth. The combination of the spot's horizontal movement and the voltage's changes up and down should result in a display that has, very roughly, a sinusoidal shape. As you continue varying the voltage quickly up and down, experiment with both vertical sensitivity and sweep time settings. Make sure that what you see makes qualitative sense. When finished, *turn off the power supply and disconnect it from the scope.*

9. A sinusoidal voltage can be produced accurately and predictably by an instrument known as a *function generator,* which you see on your bench. The function generator produces time-varying voltages (waveforms), just as the PS produces DC voltages. Just as there is a voltage control on the PS to set the magnitude of its DC voltage, there are controls on the function generator to set the amplitude of the voltage variations of the waveforms that this instrument produces. Sometimes this is done with two controls: one is continuous, and is often marked "amplitude"; the other is discrete (e.g., with three possible settings), and is often marked "attenuator." The amplitude of the variations of the waveforms is a function of the setting of *both* of these controls. There are other controls that determine how fast the variations of the waveform will be. There are also controls that determine the shape of the waveform, plus a few other controls, the function of which will become clear later on.

To prepare for use of the function generator, you will first need to set its controls as follows:

Frequency:	1 Hz (in some generators, this may necessitate setting both a dial and a multiplier button)
Amplitude:	1 V (this may necessitate setting both an *amplitude* and an *attenuator* control; if the generator does not give an indication of its amplitude, you may simply turn down the amplitude and set it after you have connected the generator to the scope below)
Function type:	sinusoidal
DC offset:	off
Sweep and modulation:	off (not all function generators have these features)

If some of these settings do not make sense for the type of generator available to you, set up the generator as specified by your lab instructor.

The scope's vertical sensitivity should be set to 1 V/div, and the scope's sweep rate should be set to 0.2 s/div.

10. Connect the output of the function generator to channel 1 of the scope. You can do this in any one of several ways, depending on the type of output connector on your generator. If that connector is coaxial, you can use a cable with a coaxial connector on each side; correct connection of grounds is then guaranteed. If the output is instead provided with "banana" receptacles, you need to use an appropriate cable or a combination of a cable and an adaptor; in this case, you need to be careful with grounds, as explained in the chapter on ground connections. Be sure that, of the two banana plugs, the one connected (through the cable) to the ground of the scope goes to the ground of the generator. (Another way to connect the scope to the generator is to use a probe again; use the $\times 1$ scope probe; if your probe has a $\times 1 / \times 10$ selector, set it to $\times 1$. The *coaxial connector* of the probe should be connected *to the scope*; the *clips* of the probe should be connected *to the function generator*. In this case, you may need to attach a short piece of wire to the nongrounded output connector of the generator so that you can clip the nongrounded tip of the probe to it.)

If you have made the connections correctly, the spot on the scope's display should be moving qualitatively as in step 8, except that now the voltage takes both positive and negative values. Vary the amplitude controls on the function generator and the sensitivity control on the scope, in a coordinated fashion. Observe the effect on the trace. Be sure you understand the function of these controls. Note that although the effect of the these controls on the scope's trace is qualitatively similar, controlling the function generator's amplitude determines how large a signal this instrument *gives*; in contrast, controlling the sensitivity of the scope determines how large the *displacement of the trace* will be per volt of the signal connected to the scope's input.

11. Vary the frequency controls on the function generator and the sweep rate control on the scope, *in a coordinated fashion, so that each time you can see a few cycles of the waveform on the screen.* Observe the effect on the trace. If necessary, adjust the intensity of the trace to obtain a convenient display. Continue experimenting, until you fully understand the function of these controls.

To prepare for the next few steps, set the generator frequency at 1 kHz, and set the sweep time on the scope so that you can observe several cycles on the screen.

12. (This step is to be done if your function generator has an *attenuation* control. If it does not, proceed to step 13.) You will now experiment with the attenuator of the function generator. This control functions like a voltage divider inside the instrument, dividing the amplitude of the waveform produced by a specified attenuation factor, which we will denote by a. For example, there may be a setting of this control corresponding to $a = 1$; this means that the entire signal appears at the output. There may be another setting corresponding to $a = 10$; this means that the amplitude of the signal appearing at the output is *one-tenth* of what it was at the $a = 1$ setting; and so on. Often, though, rather than marking the attenuation factor value a directly, what is marked is the corresponding value in decibels, or dB. You will find a discussion of decibels in Appendix C.

Try different settings for the attenuator, and verify that the attenuator markings make sense.

13. In preparation for the following step, disconnect the scope's input from the function generator. Connect a short across the scope's input. Using the scope's vertical position control, set the zero-reference level of the trace at the midpoint (see step 5). Now disconnect the short, and again connect the function generator to the scope's input. *Do not touch the scope's vertical position control, for the rest of this step and for the following step.* Set the "trigger coupling" to AC.

Turn on the *DC offset* control on the function generator. Vary its setting, and observe its effect on the trace. Why is this control called DC offset?

14. You will now experiment with the *input coupling* control on the scope (this control may or may not be marked explicitly as such). This control has settings marked DC, AC, and GND. CAUTION: There are also other controls that may be similarly marked; make sure that you have identified the one in the "channel 1" group of controls. The name of these settings may be confusing at first. Note that the names "DC" and "AC" do *not* refer to the type of voltage observed; that voltage is generated by the function generator, so only controls on *that* instrument can affect it. To find out what these names on the scope's face really mean, set the *DC offset* on the function generator to a nonzero value, and move the scope's channel 1 coupling control between the DC and AC settings. What do you observe? Repeat for a different offset value. Think of how the scope treats the DC offset of the observed waveform when the scope's coupling is set to AC. What do you believe is the function of the AC setting? In contrast, what is the function of the DC setting?

15. Change the scope's coupling control between the AC (or DC) and GND settings. Observe the effect on the scope trace. The GND setting shorts the scope's input internally to ground (without shorting the external connection), which has the same effect on the trace as if zero V had been connected across the input. This is a convenient feature, which allows you to set your zero-reference vertical level without having to go through the entire procedure outlined in step 5.

Before proceeding, turn the DC offset on the function generator off, and set the *trigger coupling* on the scope to DC.

16. Play the following educational game: Ask a lab partner to set the generator to a given amplitude and frequency, covering the generator's front panel so that you cannot see the settings. Determine the function generator settings by observing the waveform with the scope. You will need the values of both the vertical sensitivity and sweep time on the scope to do this. You can set these values as needed to obtain a convenient trace. Then, exchange roles with your lab partner and repeat.

Before proceeding, make sure you understand the function of all of the controls discussed so far.

17. You will now find out what the scope's *trigger* function does. (You should be warned right away that this function can be tricky, so you may not be able to understand it completely the first time you are using it. Further understanding will come with experience, as you use the scope in other experiments.) Use a 1 V peak, 1 kHz sinusoidal signal for this part. Set the trigger mode to *normal* (as opposed to *auto*). Choose an appropriate sweep rate so that you can observe a few cycles on the screen. Experiment with the trigger *level* and *slope* settings. You should observe that the level control determines the *value* of the voltage waveform at which the trace is triggered; that is, *starts* its left-to-right movement. The slope control determines whether the value chosen by the level control will be on an *up-going* or *down-going* part of the waveform (i.e., with a positive or negative slope, respectively).

Note that at the *normal* trigger setting, the trace is triggered only if the signal has a level and slope compatible with the *level* and *slope* settings. In contrast, at the *auto* setting, there is always a trace; when no appropriate triggering can be achieved, the trace automatically switches to a *free-running* mode. Verify this.

Triggering can also be activated by an external signal, connected to the external trigger input of the scope, by setting the trigger control to "external." It can also be activated at the frequency of the voltage on the electric power lines by setting the trigger control to "line." We will not be using these modes in this lab.

When you are finished with this part, *return the trigger controls on the scope to the settings specified in step 1.*

At this point, you have been acquainted with the main controls of the oscilloscope and the function generator. There are some other controls and features, depending on the models of the instrument. You do not need to worry about these for now; they are best studied after you have practiced, and feel comfortable with, the basic features you have just studied. More information can be found in the user's manuals for the instruments.

HEARING THE SIGNALS

18. You will now prepare to hear the waveforms produced by the function generator. *First, turn off the power supply and the function generator.* Hook up the setup shown in Fig. 3, using the power amplifier and loudspeaker (or, simply, "speaker") provided. *Use cables as short as possible, especially those that are connected to the input of the amplifier* (because long cables can act as antennas and can pick up external interference, which would interfere with the experiment).

As you see in Fig. 3, for the amplifier to operate, a power source needs to be connected to it (the power supply). Turn the voltage knob on the power supply all the way down; then turn on the power supply, and set its voltage to the value needed by the amplifier. This value will normally be indicated near the amplifier's power supply terminals; if not, ask your instructor.

19. With the function generator still off, turn its amplitude all the way down and set the frequency to 1 kHz. **Keep the speaker away from your ears, and from the ears of your partners.** Turn the generator on. Carefully and gradually, increase the generator amplitude until you can hear the tone it produces. The level should be kept low, so as to avoid disturbing the other students in the lab. Set the oscilloscope to observe the waveform as you listen to it (a few cycles should be visible on the screen). Record the peak-to-peak amplitude of the signal needed for comfortable listening; *you will need*

Fig. 3

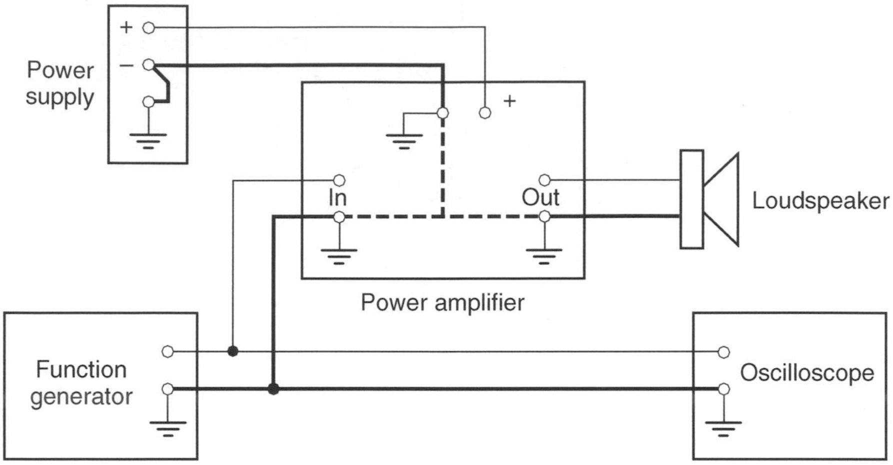

this value shortly. Experiment with the frequency control of the generator and observe the effect on the sound.

20.   Change the waveform shape control on the function generator to other (nonsinusoidal) shapes (square, triangle, etc.), and observe the waveform with the scope and its effect on the sound. In other courses, using Fourier analysis, you will be able to show that these other shapes can be represented as sums of sinusoids, with frequencies that are multiples of the basic frequency. The differences in sound that you hear can be attributed to such "harmonics."

21.   If the speaker you are using has a removable front panel, go ahead and remove it carefully. Touch the paper cone of the speaker *very lightly,* and feel its vibration for various frequencies and amplitudes. Low frequencies, from a few Hz to a couple of hundred Hz, will work best. Then, lay the speaker flat, with the paper cone facing toward the ceiling. Place a very *small* particle (e.g., the size of a grain of rice) on the cone, and observe its movement when the frequency of the signal is low (a few tens of Hz). Then remove the particle, replace the speaker panel, and set the speaker upright.

22.   Turn off the generator, and *disconnect* it from the setup. Now use the dynamic microphone provided *in place* of the generator. Set the scope's sweep time at about 2 ms/div. Increase the sensitivity of the scope so that you can observe the signal produced by the microphone when you speak into it. Clearly, although the scope is sensitive enough to show you this signal, the amplifier-loudspeaker is not sensitive enough to reproduce the sound at loud enough levels; further amplification is needed. In Experiment 5, you will design and build a preamplifier, which will increase the sound level of your setup. For now, simply observe the effect on the scope as you produce a variety of sounds in front of the microphone (speaking, whistling, clapping, etc.).

23.   What is, roughly, the peak-to-peak amplitude of the microphone signal, when you speak into it at a comfortable level and at a distance of about 5 cm (about 2″)? By

what factor would you need to amplify this signal to make it as large as the comfortable listening level you determined in step 19?

AN ASSIGNMENT FOR A DIFFERENT DAY: STUDYING THE USER'S MANUALS

You have explored the main features of the function generator and the oscilloscope in the above experiment. This is more than enough for today; you should now let what you learned sink in. In the next few days or weeks, though, you will need to return to the instruments in order to reinforce and expand what you have learned. This is the purpose of the following assignment. Your instructor may want to reserve a special lab session (or part of one) for it, or you can do it at the first available opportunity, for example at the end of a lab session in which you happen to have finished early. *But you should not neglect to do it.* Many students, due to their lack of true familiarity with measuring instruments (especially the oscilloscope), end up turning knobs at random until they get some sort of display on the screen. You can avoid this time-wasting, frustrating, and unprofessional practice by carefully doing this assignment.

Assignment

Get the user's manuals (or other description your instructor may provide) for the *specific* function generator and oscilloscope models you are using. Locate in these manuals the parts that describe the functions of each control on their panels and the use of the oscilloscope probes. The following should be done for *each* feature being described, except perhaps for some advanced features that your instructor has indicated you can skip for now.

(a) Read the description of the feature carefully, and make sure you understand it.

(b) Try the feature being described, in the manner being suggested by the description. To do so, you will need to generate an appropriate signal by using the function generator and observe it with the scope. Be sure that what you observe makes sense and that any control you are studying influences the display as you would expect from the description. Practice with the use of this control until you are confident you understand it fully.

Do not neglect to study the part that describes the probes in the oscilloscope manual. You will need to refer to the user's manuals again in the future, whenever you need to clarify some aspects of the instruments' operation.

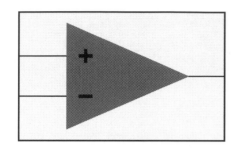

EXPERIMENT 4

BASIC CHARACTERISTICS OF OP AMPS AND COMPARATORS

Objective   In this experiment, you will work with an operational amplifier (op amp) chip. You will study its most important input-output properties. (The use of the op amp together with resistor feedback will be studied in Experiment 5.) You will then use the op amp as a simple comparator. Op amps are widely used in electronics. Understanding them will take you a long way toward being able to design and build useful and interesting circuits, including several in this book.

Preparation   Op amps are almost certainly discussed in the theory text you are using. To prepare for this experiment, you need only study the introductory part in your text, describing the input-output characteristics of the op amp by itself (i.e., not connected to any other elements). If your text does not cover op amps, do not worry: all you need to know about them in order to do this experiment will be explained along the way.

MAKING A BIPOLAR POWER SUPPLY

1.   In this experiment, you will need a way to generate a voltage that can vary continuously around zero. Most power supplies cannot do this. Fig. 1 shows a way to obtain both positive and negative values for a voltage v, depending on the potentiometer setting. Voltages V_1 and V_2 can be obtained conveniently from the two outputs of a dual power supply. The element labeled C is a capacitor, with a value of 33 nF, used to reduce the possibility of parasitic interference. (The wires you use to connect your circuits can, in some cases, act as antennas for interference, e.g., from a radio station or from the lights and the instruments in the room. The capacitor helps make the voltage v constant, i.e., less susceptible to such interference. There are several capacitors inside your power supplies, too. You will work with capacitors in Experiment 6. With DC voltages, capacitors act as open circuits, so you can ignore the presence of the capacitor, although you should connect it as shown.)

Fig. 1

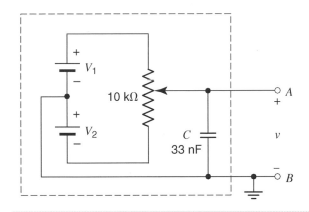

In Fig. 1, the voltage v can take any value between V_1 and $-V_2$, depending on the setting of the potentiometer. Explain this fact, and verify it experimentally.

Note that if you set the two supplies at very small voltage values, the potentiometer allows you extra resolution that would not be obtainable from the supplies directly. To verify this, set V_1 and V_2 so that v can be varied from about -20 mV to about $+20$ mV as the potentiometer is rotated from one extreme to the other.

Do **not** disassemble this circuit; you will need it shortly.

OP AMPS

The op amp is an active element that needs to be supplied with power to operate. A common way to supply this power is shown in Fig. 2(a). Two power supply voltages are used, with equal values denoted by V_{CC} (often in the range of 5 V to 15 V). *Notice the polarity of each supply voltage.* The common node between the supplies is the ground node. The op amp's output voltage is taken between the output terminal and the ground node. The remaining two terminals are the input of the op amp.

We will follow common practice and draw the circuit of Fig. 2(a) as shown in Fig. 2(b). Here, the power supply voltages are *understood* to be connected at the appropriate terminals, although the supply voltage sources are not explicitly shown. Sometimes, even the supply connections shown in Fig. 2(b) are omitted from circuit diagrams, but they are always understood to be there. *Remember*: Whether the power supply connections are explicitly shown or not, power must be properly connected to an op amp before it can operate.

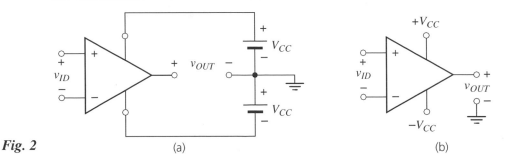

Fig. 2   (a)   (b)

Fig. 3

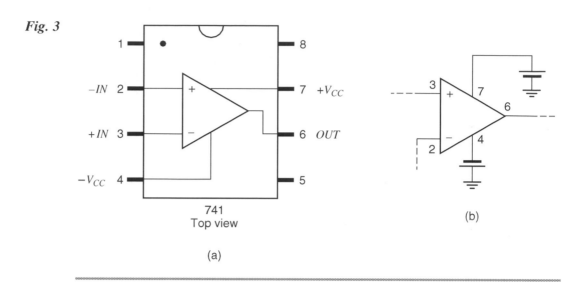

741
Top view

(a)

(b)

In this experiment, we will use a popular op amp type known as the 741. It comes in a package, with metal pins that can be pushed into a prototyping board such as the one you will be using for this experiment (for a more permanent connection, the pins can be soldered to a printed circuit board). A widely used package for this op amp, when viewed from above, is shown in Fig. 3(a). With the package viewed from above and positioned as shown (mark at the top), the pin numbering is understood to be as indicated. The pins that correspond to the input, output, and power supply connections are indicated. It is often convenient to transfer the pin numbers onto circuit diagrams, as shown, for example, in Fig. 3(b). You can do this on the diagrams that follow if you find this practice convenient. We will only be using the pins shown; the rest of the pins should be left unconnected for our experiment (some of these pins do have a function but not one that is essential for the purposes of this experiment). If, for the 741 chips you will be using, the package (and associated pin numbering) is different from that shown in Fig. 3(a), your instructor will supply you with information for that package.

When the op amp is connected to other elements, extremely small currents flow at its input terminals. For our purposes, *these currents can be assumed to be zero.* The most basic function of the op amp is the following: If the voltage across its input (v_{ID} in Fig. 2) is very small, the output voltage (v_{OUT} in Fig. 2) is a very large multiple of the input. However, the output cannot be outside the voltage range provided by the power supply; in other words, the large voltage gain just mentioned is possible only if the input does not demand that the output be outside the supply voltage range; otherwise, the output saturates at a value close to one of the supply voltages. The behavior just described is shown by the curve in Fig. 4. In the next part, you will determine this curve for an op amp chip provided by your instructor.

OP AMP DC TRANSFER CHARACTERISTIC

2. Set up the circuit shown in Fig. 5. The two power supply voltages for the op amp can conveniently be obtained at the two outputs of a dual power supply. (You are already using a dual power supply in the circuit of Fig. 1; use *another* dual power supply to power the op amp.) The op amp's input voltage v_{ID} is provided by the circuit of Fig. 1

Fig. 4

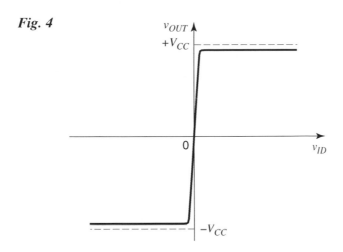

Fig. 5

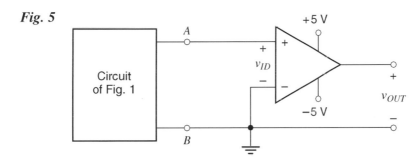

as shown. The various ground connections may be confusing in the beginning. If so, refer to the chapter on ground connections. Remember, all points indicated by the ground symbol should be connected together.

3. To obtain the input-output DC characteristic of the op amp, you need to feed various voltage values at its input (v_{ID}) and measure the corresponding voltage at its output (v_{OUT}). First, obtain a coarse characteristic by measuring and plotting v_{OUT} as a function of v_{ID}, for v_{ID} ranging from –2 V to +2 V. Is the plot what you expected?

4. Now obtain a fine characteristic, showing the details when v_{ID} is varied over a range of 1 mV or so *around the point that the output crosses zero*. To obtain high resolution for v_{ID}, you will need to set V_1 and V_2 (see Fig. 1) at *very* small values (e.g., 10 mV or less). This may be difficult to achieve, but try your best.

5. What is, approximately, the value of v_{ID} at which the output crosses 0? This value is called the *equivalent input DC offset voltage* (or simply *input offset*) of the op amp, and it should ideally be zero; in practice, it can be several mV and can vary with temperature, as well as from chip to chip.[1]

---

[1]With the 741 op amp, the input offset can be nulled externally if desired. This can be done by connecting the main body of a 10 kΩ potentiometer across pins 1 and 5. The slider of the potentiometer is connected to the negative supply voltage. The potentiometer can then be adjusted until the input offset is nearly zero. We will not be doing this here.

Fig. 6

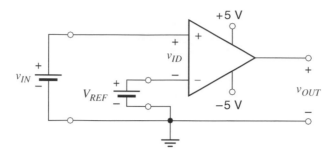

6.   Try to obtain an approximate value for the *slope* of the fine plot, *in the region where this slope is high.* You will find, with our simple setup, that determining this slope is from very difficult to impossible (it is instructive, nevertheless). Just do your best to provide a rough estimate. The value of this slope is the *DC gain* of the op amp.

  When you are finished with this part, turn off the circuit and disassemble the circuit of Fig. 1.

THE OP AMP AS A COMPARATOR

Because of the shape of its coarse transfer characteristic, which you obtained in step 3, the op amp can be used as a comparator. For positive v_{ID} values larger than several mV, the output is high (close to $+ V_{CC}$); for negative v_{ID} values with magnitude larger than several mV, the output is low (close to $- V_{CC}$). In other words, the comparator compares its input voltage to (approximately) 0 V and gives a corresponding output indication ("high" or "low"). The circuit to the right of terminals A and B in Fig. 5 was just such a comparator.

7.   Consider now the circuit of Fig. 6 (do *not* build it yet). Define v_{ID} again as the voltage of the + input terminal of the op amp *with respect to its – input terminal.* (Note: this terminal is *no* longer grounded.) What is the expression of v_{ID} in terms of v_{IN} and V_{REF}? Using this result and your coarse plot from step 3, try to predict what the value of v_{OUT} will be if (a) $v_{IN} > V_{REF}$ and (b) $v_{IN} < V_{REF}$.

8.   Now build the circuit in Fig. 6. Set the reference voltage V_{REF} to a fixed value of no more than 2 or 3 volts, and raise v_{IN} slowly from 0 past the value of V_{REF}, observing the value of v_{OUT}. Is your observation consistent with your expectations from step 7? Now set V_{REF} to a different value, and repeat. This circuit is a comparator; state clearly *what* it compares to *what.*

OBTAINING A VISUAL INDICATION

Note that for comparison, the exact value of v_{OUT} is not important; all that matters is whether the value is "high" or "low." You can indicate these two states of the output in various ways, for example, by turning on a light when the output is high and by turning it off when the output is low. You will now do this, using a *light-emitting diode,* or LED.

Fig. 7

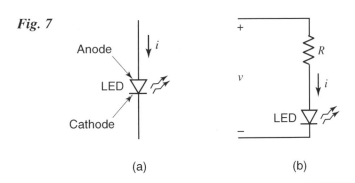

(a)   (b)

The symbol for an LED is shown in Fig. 7(a). When a sufficient amount of positive voltage is applied in the forward direction (from the *anode* to the *cathode*), the LED emits light. However, one should *not* use an LED directly across a voltage source since if the voltage happens to be even a little larger than a maximum allowable value, the LED current can be excessive, and the device can be damaged. To avoid this possibility, one can use the LED in series with a resistor, as shown in Fig. 7(b). If the total voltage across the combination is v, part of this voltage will appear across the resistor, and thus the resistor current will be less than v/R. This current is the same as the LED current, and it is easy to limit its value by choosing a suitable value for R. A convenient value for our purposes is 1 kΩ (a different value may have to be used, depending on the type of LED and the total voltage applied across the resistor-LED combination). If the voltage v in Fig. 7(b) is positive, current will pass through the LED in the forward direction, and the LED will light it up; if v is instead negative, the LED will be off.[2]

Any common LED will do for this experiment, as long as it provides a sufficiently bright display. To distinguish the two terminals of the LED, some manufacturers flatten its plastic casing on the cathode side or make the anode terminal longer. Other ways are used, too. If you are not sure which terminal is which and the LED does not light up, switch its terminals and try again.

9.   Connect the circuit of Fig. 7(b) to the circuit of Fig. 6 in such a way that when $v_{IN} > V_{REF}$, the LED is on; and when $v_{IN} < V_{REF}$, the LED is off. Verify this operation.

10. Modify the circuit in the previous step so that when $v_{IN} > V_{REF}$, the LED is *off*; and when $v_{IN} < V_{REF}$, the LED is *on*. Verify this operation.

     When you are finished with this step, disconnect v_{IN} and the resistor-LED combination from the rest of the circuit.

---

[2] When a LED is on, a typical value for its forward voltage (i.e., of the anode with respect to the cathode) is 1.7 V. If a voltage is applied in the reverse direction (i.e., for the cathode to the anode), its value must not exceed a "breakdown" value specified by the manufacturer (typically 5 V).

Fig. 8

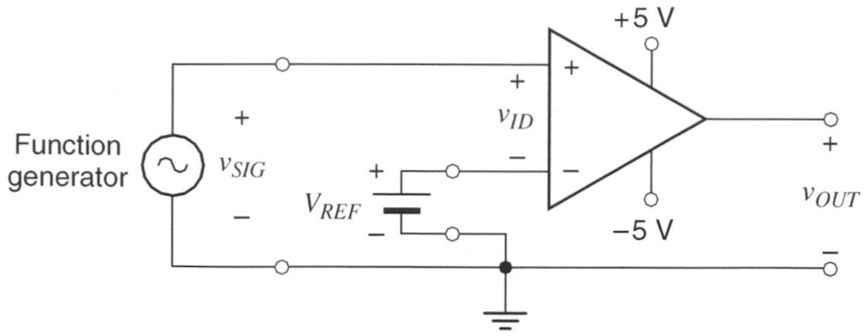

THE COMPARATOR WITH AN AC INPUT

11. Now assume that the DC source v_{IN} is replaced by the function generator, as shown in Fig. 8 (do *not* build this circuit yet). Can you *guess* what the shape of the output waveform will be if the generator signal v_{SIG} is a sinusoidal voltage of peak amplitude 1 V and the reference voltage V_{REF} is 0.5 V?

12. Now connect the circuit of Fig. 8. Use a sinusoidal voltage v_{SIG} with a frequency of 100 Hz and an amplitude of 1 V. Make sure that the DC offset *of the function generator* is zero (this quantity has *nothing* to do with the equivalent input DC offset of the op amp mentioned above). Set V_{REF} to 0.5 V. Observe the input on channel 1 of the scope, and the output on channel 2.[3] The input coupling of both channels should be set to DC. The triggering should be set to channel 1. Be sure that you trigger correctly so that you can obtain a stable display. Is what you see as expected from step 11?

13. Try different values for the amplitude of the sinusoidal signal, as well as for the DC voltage V_{REF} (do *not* raise the value of these quantities over 2 or 3 volts). Make sure you can explain what you see. Try waveforms other than sinusoidal for the signal, and again be sure you understand what you see.

OP AMP SPEED LIMITATIONS

14. Set V_{REF} to 0. For v_{SIG}, use a sinusoidal waveform with an amplitude of 1 V. Increase the frequency of the signal above 100 Hz; as you do so, make sure you adjust the sweep rate on the scope so that the display always shows a few cycles of the signal. Does the circuit function properly as a comparator at very high frequencies? If you think there is a problem in this respect, describe it. (The problem you see is caused by the internal circuitry of the op amp; an explanation of this is beyond the scope of this lab. Although it is convenient for us to use an op amp as a comparator, this is not the main function op amps have been designed to perform, so do not be surprised by what you see. Other chips, designed specifically to be comparators, can do a better job with comparisons at high frequencies.)

---

[3]The setup you will need on the oscilloscope to observe both channels will depend on the particular instrument you are using. On some oscilloscopes, there may be a special setting for such "dual trace" operations. On analog oscilloscopes, the vertical mode should be set to CHOP (the ALT setting is more appropriate at faster sweep rates).

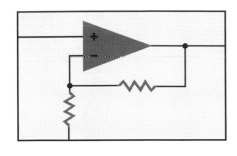

EXPERIMENT 5

AMPLIFIER DESIGN USING OP AMPS; A SOUND SYSTEM

Objective

In this experiment, you will design a preamplifier for your microphone, using an op amp. You will use this preamplifier in a complete sound system.

Preparation

Read about the use of op amps and resistors to make amplifiers in your textbook. A simple introduction to this subject is also provided in the following section.

BACKGROUND

In contrast to the comparators studied in Experiment 4, which provided outputs equal to one of the two saturation values (corresponding to the flat parts in Fig. 4 of that experiment), amplifier circuits work correctly *only* if the output voltage of the op amp is *not* high enough to reach saturation; in other words, *the op amp must be operating in its high-gain region* (the region of high slope in Fig. 4 of Experiment 4). Such an operation will be assumed in the following discussion unless otherwise indicated.

Consider the circuit of Fig. 1. Resistors R_1 and R_2 form a voltage divider fed from the output voltage v_{OUT}, which develops a fraction of that voltage as v_1. From the voltage divider equation, this voltage is given by:

$$v_1 = v_{OUT} \times \frac{R_1}{R_1 + R_2} \qquad (1)$$

If, when the power supplies are first turned on, v_1 happens to be less than v_{IN}, the difference $v_{ID} = v_{IN} - v_1$ will be positive and will thus cause an increase in v_{OUT} (see Fig. 4 in Experiment 4). Through the voltage divider, this will in turn increase v_1 toward the value of v_{IN}. If, instead, v_1 starts at a value larger than v_{IN}, the difference $v_{ID} = v_{IN} - v_1$ will be negative and will cause an decrease in v_{OUT} (again, see Fig. 4 in Experiment 4). This in turn will decrease v_1 toward the value of v_{IN}. Thus, in either case, this *feedback* action will continue to change v_1 until v_1 is very close to v_{IN}. For op amps with large voltage gain, the slightest difference $v_{IN} - v_1$ activates

Fig. 1

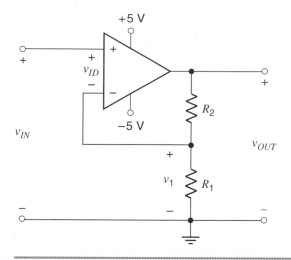

this feedback action, and v_1 becomes almost equal to v_{IN}. Thus, equating v_1 from Eq. 1 to v_{IN}, we have

$$v_{OUT} \times \frac{R_1}{R_1 + R_2} = v_{IN} \qquad (2)$$

and solving for the ratio v_{OUT} to v_{IN}, we find

$$\frac{v_{OUT}}{v_{IN}} = 1 + \frac{R_2}{R_1} \qquad (3)$$

In other words, the circuit in Fig. 1 acts as an amplifier with voltage gain v_{OUT}/v_{IN} as given above. This gain can be set by choosing appropriate values for R_1 and R_2. Because the algebraic sign of the voltage gain is positive, the amplifier is said to be noninverting.[1]

DESIGNING AND TESTING A VOLTAGE AMPLIFIER

1. You will recall from Experiment 3 that the signal produced by the microphone was not enough to drive the power amplifier at a reasonable level. You determined at that time that you would need to amplify the microphone signal by a large factor. Design the circuit of Fig. 1 for this purpose, so that it gives you an amplification factor (voltage gain) of approximately 100. Use $R_1 = 1$ kΩ. Select R_2 so that the gain magnitude is approximately as desired. Use a 741 op amp. The pin assignment for this op amp has been given in Fig. 3(a) of Experiment 4. Given the tolerance of the resistors you are using (Appendix A), what is the minimum and maximum voltage gain that can be expected from this circuit?

---

[1]The circuit of Fig. 1 is sometimes drawn in a different, but equivalent, way. If this is the case in your theory textbook, make sure you recognize the equivalence of the circuit given there for a two-resistor, non-inverting amplifier to the circuit in Fig. 1.

Do not use your circuit to amplify the microphone signal yet. Rather, first investigate the circuit as explained in the next step.

2. Measure and plot the DC transfer characteristic (v_{OUT} vs. v_{IN}) for the circuit of Fig. 1. Make sure you have enough points in the high-slope region so that you can obtain a relatively accurate plot. If the characteristic does not pass exactly through the origin, do not worry (this is due to an op amp imperfection called DC offset voltage, discussed in Experiment 4; although this imperfection can cause problems in certain cases if not properly taken into account, it will not cause problems in the circuits we will be using). Determine the slope of this characteristic in the high-slope region, and verify that it is equal to the desired voltage gain.

    When finished with this step, turn off the power supplies and disconnect the input source v_{IN} from the circuit.

3. By only looking at the plot obtained in the previous step (not by experimenting), answer the following questions:

    (a) What is the input voltage range over which the circuit behaves linearly?

    (b) What is the maximum and minimum output voltage possible?

    (c) If the input voltage were a sinusoid, what would its maximum amplitude be before it would drive the circuit into its nonlinear region?

    (d) What would the shape of the output voltage be if that amplitude were *not* exceeded?

    (e) What would be the shape of the output voltage if that amplitude *were* exceeded?

4. To test your preamplifier, drive its input with a 1 kHz sinusoidal signal produced by the function generator. Observe both v_{IN} and v_{OUT} simultaneously, using the two channels of the scope.[2] Try different amplitudes, and verify your predictions in step 3. Is the input amplified by the expected factor? If not, debug the circuit. If necessary, ask your lab instructor for help.

5. Replace the function generator by the dynamic microphone. Using the two channels of the scope, verify approximately that the microphone signal is amplified by the desired amount.

A SOUND SYSTEM

6. You are now ready to use the microphone preamplifier you just designed in a complete sound system. The microphone signal has already been fed to the input of the preamp so that the latter can amplify it and develop a multiple of it at its output. This

---

[2]If you are using a digital scope, consult the manual on how to achieve dual-channel operation. On analog scopes, you should set the vertical mode to CHOP and make sure you trigger appropriately, so that the waveforms on the display are stable.

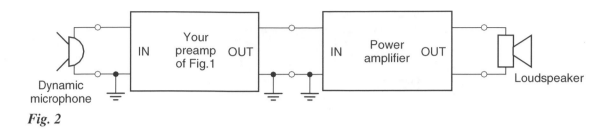

Fig. 2

amplified output must now be fed into the power amplifier, which in turn should drive the speaker. These tasks can be performed by the circuit of Fig. 2. Power supply connections are not shown but are *understood* to be there, for *both* the preamplifier *and* the amplifier. Set this circuit up, *with all power supplies off.* This is the largest circuit you have built so far in this lab, so the whole thing may be somewhat confusing at first. *Avoid using unnecessarily long cables,* and keep everything neat, as a circuit of this size can become unwieldy. Pay special attention to the power supply and ground connections. If in doubt, consult the chapter on ground connections. Carefully inspect the circuit, making sure that everything is connected correctly.

7.   **CAUTION: Keep the speaker away from your ears, and from the ears of others.** Keep the microphone away from the speaker. Turn on the power to the two amplifiers. Try out the sound system. If you hear whistlings, move the microphone further away from the speaker. Speak loudly into the microphone, but not loudly enough to distort the sound coming from the loudspeaker. If you find that the gain is not enough, you can increase it by increasing R_2 in Fig. 1 (be sure you turn off all power before you do so).

8.   Use the scope to observe the signal at the input of the preamp, at the output of the preamp, and across the speaker. Be sure that you connect the ground terminal of the scope probe to the ground of the circuit. You can compare any two of these signals by displaying them simultaneously on the two channels of the scope. Speak loudly, until you hear distortion in the sound, while you watch the signal across the loudspeaker. What do you observe in the waveforms when you hear distortion?

Notice that it is your preamp, rather than the power amp, that amplifies the signal voltage. However, the power amp serves a very important function: It provides the large current needed by the speaker. Your preamp could not do this. If the speaker were fed directly from the output of the preamp, the system would not work well. The way the system of Fig. 2 works is as follows: The preamp amplifies the microphone voltage signal and drives the amplified voltage into the input of the power amp. This input behaves as a large resistance, and thus the signal current it draws from the preamp is small. The output of the preamp can provide this current. The power amp then passes the signal to the speaker through a power stage, which can supply the large signal current and power demanded by the speaker. The way in which this is accomplished inside the power amplifier is studied in electronics theory courses.

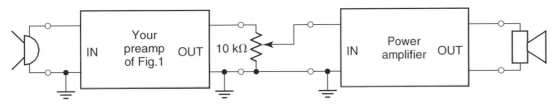

Fig. 3

VOLUME CONTROL

9. You will now add a *volume control* to your sound system. To do so, you need to take the output of the preamplifier, and *divide down* its voltage by a factor that can be adjustable. The resulting voltage should then be fed to the input of the power amp. You can do this by using a 10 kΩ potentiometer as a variable voltage divider, as shown in Fig. 3. (A 100 kΩ potentiometer is also acceptable; why is the exact value of the potentiometer resistance not critical?) Try your new system by speaking into the microphone and varying the potentiometer. Does your volume control work?

10. Increase the volume to maximum (**while always keeping your ears away from the speaker**), and move the microphone close to the speaker. What do you observe? Can you explain it?

REVERSING A TRANSDUCER ACTION

11. The loudspeaker converts electricity to sound. Could the loudspeaker perhaps be used to convert sound to electricity? That is, could it act as a microphone? Borrow a speaker from the group next to you, and try to find out. Leave the speaker on their bench, and connect it to your system with long wires. Be sure that the two speakers are very far away from each other, to avoid the effect you observed in part 10.

   CAUTION: This step deals with the use of the *speaker* as a *mike*. Do not try to use the *mike* as a *speaker*, as the power delivered to it can permanently damage it.

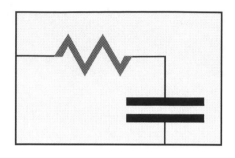

EXPERIMENT 6

RC CIRCUIT TRANSIENTS; MORE ON MEASUREMENT TECHNIQUES

Objective

In this experiment you will study circuits that use one capacitor and resistors. You will observe the charging and discharging of the capacitor in such circuits, and you will measure the associated time constant. This experiment is also convenient for introducing certain issues associated with measurement errors and techniques. Thus, you will study how the effect of finite voltmeter resistance can affect your circuit and how it can be properly taken into account. You will also learn how to perform floating voltage measurements with the oscilloscope.

Preparation

Read about first-order RC circuits in your circuit analysis text.

BACKGROUND

The phenomena encountered in resistor-capacitor (RC) circuits are of fundamental importance in electrical engineering. They are found in every piece of electronic equipment, either intentionally, as in some filter circuits, or unintentionally, as conductors have both parasitic capacitance to their neighbors and resistance. In fact, this is one of the main reasons that the speed of a computer is limited. To prepare for this experiment, you should study the material on simple RC circuits in your textbook. For convenience, the main conclusions of this material are summarized below.

Consider the circuit of Fig. 1(a). Assume that the capacitor is initially discharged ($v_C = 0$). If, at time $t = 0$, the switch is thrown to position 1, the voltage $v_R = V_{PS} - v_C$ across the resistor causes a current v_R/R through it in the direction shown, which deposits positive charges on the top plate of the capacitor. Thus the voltage of the capacitor, v_C, increases. Initially, the charging current is large, so the voltage of the capacitor rises fast; as it rises, though, the voltage across the resistor,

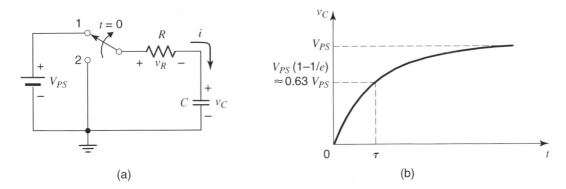

Fig. 1

$V_{PS} - v_C$, decreases, and therefore the resistor current decreases, too. Thus charges are now delivered to the capacitor plate at a slower rate, and the capacitor voltage does not rise as fast as before. This process continues as long as there is current in the resistor to charge the capacitor. It can stop only when the resistor voltage (and thus its current) becomes zero, that is, when $v_R = V_{PS} - v_C = 0$. This means that when the charging stops, we must have $v_C = V_{PS}$. A little thought will show that this situation will only be approached asymptotically, as indicated in Fig. 1(b). In your textbook it is proven that the capacitor voltage obeys the relation

$$v_C(t) = V_{PS}\left(1 - e^{-\frac{t}{\tau}}\right) \tag{1}$$

with

$$\tau = RC \tag{2}$$

being the time constant of the circuit. The plot of Eq. 1 can easily be seen to be in the form shown in Fig. 1(b). It can be seen from the equation that when $t = \tau$, the voltage has reached the value $V_{PS}(1 - 1/e)$, which is about 63% of the final value V_{PS}. A few time constants later, the capacitor is almost completely charged.

Assume now that the capacitor has been fully charged to the value V_{PS}, and at $t = 0$ the switch is thrown to position 2, as shown in Fig. 2(a). Now the entire capacitor voltage v_C is applied across the resistor, and it causes a current v_C/R in the direction shown. This current *removes* positive charges from the top plate of the capacitor, and thus v_C decreases. As v_C gradually becomes smaller, the resistor current v_C/R becomes smaller, too; thus, it removes charges from the capacitor at a slower rate, and v_C changes more slowly than before. The capacitor voltage will stop changing only when the current that discharges it, v_C/R, becomes zero, which means that the discharging will stop when $v_C = 0$. This will occur asymptotically, as indicated in Fig. 2(b). In your theory textbook, it is proven that the capacitor voltage follows the equation

$$v_C(t) = V_{PS}\,e^{-\frac{t}{\tau}} \tag{3}$$

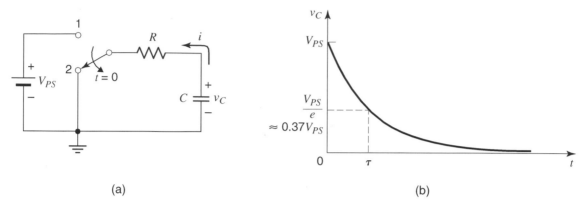

(a)   (b)

Fig. 2

which is of the form of the plot in Fig. 2(b). From this equation we see that at $t = \tau$, the capacitor voltage has dropped to V_{PS}/e, that is, to about 37% of its initial value of V_{PS}.

Effect of voltmeter resistance

To observe the above effects, we will use a voltmeter across the capacitor, as shown in Fig. 3. Unfortunately, the voltmeter is not perfect; its input is equivalent to a resistance, which we will denote by R_{VM}. A common value for this resistance is 10 MΩ. When the voltmeter is attached across two points, its input resistance is effectively connected across these points. In some cases, as in this experiment, one must be aware of this effect and take it into account.

Assume that the switch is in position 2. Then the circuit of Fig. 3 is equivalent to that in Fig. 4(a). If the voltmeter is replaced by its equivalent resistance, we obtain the circuit of Fig. 4(b). Note that the two resistances in this circuit are in parallel; combining them, we obtain the circuit of Fig. 4(c), where

$$R = \frac{R_1 \times R_{VM}}{R_1 + R_{VM}} \tag{4}$$

Thus, the capacitor discharges through a resistance that is *not* equal to R_1, as one might have concluded by carelessly looking at Fig. 3, but rather through a resistance equal to R. It is clear from Eq. 4 that the effect of the voltmeter resistance can be ignored only if $R_{VM} \gg R_1$, in which case we would obtain $R \approx R_1$. In general, we try to use voltage-

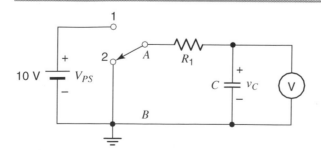

Fig. 3

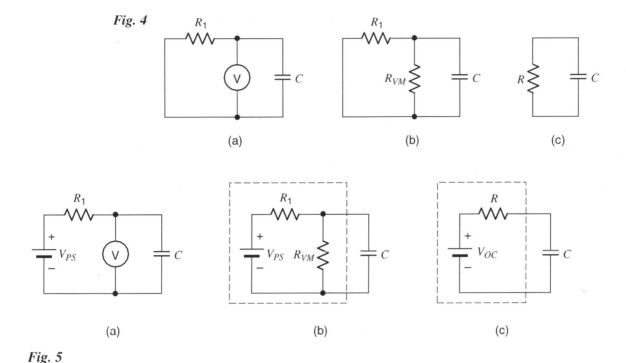

Fig. 4

(a)   (b)   (c)

(a)   (b)   (c)

Fig. 5

measuring instruments (such as voltmeters or scopes) with input resistance as large as possible, ideally much larger than the effective resistance of the circuits we are trying to measure.

Assume now that the switch in Fig. 3 is in position 1. Then the circuit is as shown in Fig. 5(a). Replacing the voltmeter by its internal resistance, we obtain the circuit of Fig. 5(b). Consider now the part of the circuit inside the broken line. This part can be replaced by the equivalent circuit shown in Fig. 5(c). In circuit theory, this is known as a Thevenin equivalent circuit. The way to calculate the values of V_{OC} and R in this circuit is explained in your circuits textbook. It can be shown that R can be found by replacing V_{PS} in Fig. 5(b) by a short, and by looking at the resulting resistor combination; since the two resistors will then be in parallel, R is given again by Eq. 4. The value of V_{OC} can be found by calculating the open-circuit voltage across the output of the broken-line box in Fig. 5(b), with the capacitor removed. This, using the voltage divider equation, gives

$$V_{OC} = V_{PS} \times \frac{R_{VM}}{R_1 + R_{VM}} \tag{5}$$

Examining Fig. 5(c), we can see that the capacitor charges to a final voltage of V_{OC} (*not* V_{PS}), at a speed determined by C and R (*not* R_1). Similar conclusions would have been reached if, instead of a voltmeter, we had used an oscilloscope with finite input resistance.

In preparation for this experiment, verify the above formulas and calculate R and V_{OC} for $R_1 = 10$ MΩ, $R_{VM} = 10$ MΩ, and $V_{PS} = 10$ V.

CAUTION: Capacitors of the electrolytic type, usually with values 0.1 μF or above, have polarity indicated on them, which specifies which terminal is to be

connected to the most negative (or positive) potential. If this polarity is not observed, such capacitors can malfunction or even explode. Be sure that capacitor polarity is always observed. If in doubt, ask your instructor.

RC CIRCUIT TRANSIENT RESPONSE

1. Construct the circuit shown in Fig. 3. Use the single-pole, double-throw switch provided (before connecting this switch, identify its leads by using the ohmmeter, as in step 10 of Experiment 2). Use $V_{PS} = 10$ V, $R_1 = 10$ MΩ and C = 10 μF. **If a polarity is indicated on the capacitor you are using, make sure to observe it.** The switch should be in position 2 until you are ready to start. The capacitor voltage should be monitored with the DMM, which initially should show that the capacitor voltage is zero. If that voltage is not zero, discharge the capacitor by momentarily shorting it with a piece of wire.

2. Throw the switch to position 1, and use a watch to record the capacitor voltage, v_C, versus time, t (in s). Define the moment you throw the switch to position 1 as $t = 0$. Take data for about 5 min. If you make a mistake and you need to start over, make sure you discharge the capacitor first by momentarily shorting it with a piece of wire.

3. Plot v_C versus t, using the data obtained in step 2. Is the final value of the voltage what you expect? Why does it take so long for the capacitor to charge up?

4. Now move the switch to position 2 in order to observe the discharging of the capacitor. Again record the capacitor voltage, v_C, versus time, t.

5. Plot v_C versus t, using the data obtained in step 4. From this plot, deduce the discharge time constant (which is equal to the time it takes for the voltage to decay to $1/e \approx 0.37$ of its initial value). Calculate accurately the expected time constant and compare it with the one obtained from the plot. Take into account the information provided under "Background".

6. Try out smaller values for R_1 or C, and observe their effect on how fast the circuit charges or discharges.[1] You do not need to produce plots for this part, but report qualitatively what you observe.

OBSERVING THE TRANSIENT RESPONSE WITH THE SCOPE

For the next step, we will decrease the value of R_1 to 270 kΩ, which will make the time constant small. Now you will no longer be able to use a voltmeter and a watch to observe the behavior of the circuit, so you will be using the oscilloscope instead. The scope, together with the probe, exhibits an effective resistance between the

---

[1]Reading values off capacitors can be confusing; some guidelines are provided in Appendix A.

Fig. 6

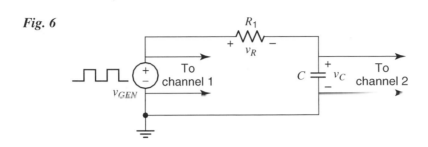

probe's tip and ground. The input resistance of the $\times 1$ probe is not much larger than the value of R_1 (it is often only 1 MΩ), and this can really affect the behavior of the circuit, as already discussed. To avoid this, you will use a $\times 10$ probe, which has a larger input resistance (often 10 MΩ). However, keep in mind that a $\times 10$ probe attenuates the signal 10 times, and *this must be taken into account when reading voltage values off the scope screen*. In addition, when using probes with time-varying waveforms, they must be *calibrated* first. Ask your instructor whether your probe is already precalibrated for you; if it isn't, you can calibrate it as outlined in Appendix B.

7. Disconnect the DMM from the circuit, and replace it with the scope. Use channel 1 for the voltage across points A and B in Fig. 3, and channel 2 for the capacitor voltage. Make sure you connect ground to ground, as usual. Try $R_1 = 270$ kΩ, $C = 1$ μF, and a sweep rate of 0.2 s/div on the scope. The input coupling of the scope should be set at DC. The triggering should be at AUTO. Throw the switch back and forth as the trace moves. You should see that the trace for channel 2 follows a plot of capacitor voltage versus time, resembling the ones you produced by hand above.

8. Try smaller time constants, each time using an appropriate setting for the sweep rate on the scope for convenient observation.

9. Think of the voltage waveform produced between points A and B when you throw the switch back and forth. What is its shape?

THE RC CIRCUIT WITH PERIODIC EXCITATION

10. The above waveform of the voltage between points A and B can be generated automatically (and at a much faster repetition rate) by replacing the voltage source/switch/hand combination by a square voltage waveform, which can be produced by the function generator. To try this, set up the circuit of Fig. 6. To observe what happens using the scope, connect the latter's channel 1 across the generator output, and channel 2 across the capacitor. Be *very* careful with the ground connections; otherwise, *you may damage the generator!* If unsure, consult the chapter on ground connections. Set the scope's trigger to channel 1 and the triggering slope to positive. Observe the generator and capacitor voltage waveforms for the same R_1 and C values as in Step 7 and for a repetition rate of 0.5 Hz for the generator waveform. If necessary adjust this rate *and the sweep rate on the scope* so that you can display on the screen two complete cycles, with the charging and discharging in each cycle being practically complete.

11. Repeat for $R_1 = 16$ kΩ and $C = 10$ nF. Change the scope sweep rate setting and the function generator repetition rate so that you can display a couple of practically complete charge/discharge cycles. Use the scope display to determine the time constant, and compare it to the value you expect.

12. Given the shape of the generator voltage (square) and the observed shape of the capacitor voltage, can you predict what the waveform of the voltage v_R across the resistor (see Fig. 6) should look like?

MEASURING A FLOATING VOLTAGE WITH THE SCOPE

13. To verify your prediction in the previous step, you need to display the waveform for v_R. *However*, you *cannot* connect a probe's input across the resistor. If you did so, where would the ground clip of the probe be attached? If it were attached to the resistor terminal on the right, it would place a ground at that point, and then the capacitor would be shorted out since its bottom terminal is also at ground (see also Fig. 2 of Experiment 3 and the associated discussion). If, instead, the probe's ground clip were attached to the resistor terminal on the left, it would short out the function generator, which would be even worse, as damage could be caused to the generator. Voltages like v_R, defined across two terminals, neither of which is grounded, are sometimes referred to as floating voltages.

The way to display v_R is to leave the probes where they were in step 11 and to realize that, from Kirchhoff's voltage law, $v_R = v_{GEN} - v_C$. Thus, all you need to do is subtract the waveform being displayed on channel 2 (v_C) from the waveform displayed on channel 1 (v_{GEN}). Most scopes provide for this operation. On some scopes, you can simply select the operation channel 1 – channel 2. Others provide an invert function for one of the channels (here assumed to be channel 2). When this control is activated, the display for channel 2 will show the negative of the waveform applied to channel 2; that is, it will display $-v_C$. On these scopes, you should also be able to select the function ADD, which would add v_{GEN} to $-v_C$. In either case, you need to make sure that the sensitivity (V/div) *is the same for both channels* (and to turn both sensitivity controls to the calibrated, or CAL, position).

Now observe the waveform for v_R, and compare it to your prediction in step 12.

EXPERIMENT 7

FILTERS, FREQUENCY RESPONSE, AND TONE CONTROL

Objectives

In this experiment, you will assemble and investigate circuits that show a preference for certain signal frequencies: The amplitude of their output signal depends not only on the amplitude of the input signal but also *on its frequency*. You will study the behavior of such "filter" circuits at various frequencies and display it in the form of plots known as frequency response. You will use filters to implement a tone control for your sound system.

BACKGROUND

Low-pass filter

You may have already been introduced to the concept of filtering in your theory class; if so, review the material on the frequency response of simple RC circuits. But even if you have not seen such material yet, you will be able to follow the intuitive explanations provided here.

　　Consider the circuit of Fig. 1. In Experiment 6, you saw that if a square wave with a large period is applied to its input, the capacitor has the time to charge or discharge almost fully during each half interval, as shown in Fig. 2(a). The output then follows the input rather closely. If, however, the period of the input is made small, the output capacitor does not have time to charge or discharge fully, and the situation illustrated in Fig. 2(b) is observed. Now the output cannot follow the large excursions of the input, and its excursions become small.

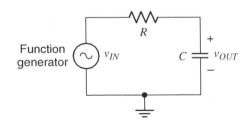

Fig. 1

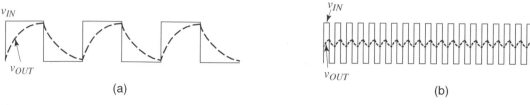

(a)   (b)

Fig. 2

These observations do not hold only for square wave inputs. Other shapes, such as triangle or sinusoidal, show qualitatively the same behavior. This is illustrated for a sinusoidal input in Fig. 3. As seen, low input frequencies allow the output to virtually follow the input, as seen in Fig. 3(a). We say that the low-frequency input signal "passes" to the output. On the other hand, when the input frequency is high, the output cannot follow the input, as shown in Fig. 3(b), and the output amplitude becomes small. The input signal, to a significant degree, is rejected, or "does not pass," to the output. For these reasons, the circuit is said to be a *low-pass filter*. Many different circuits are low-pass filters, but we will concentrate on the simple one in Fig. 1 and study it in detail.

Because at very low frequencies the capacitor voltage is almost equal to the input voltage, as shown in Fig. 3(a), we say that "at very low frequencies, the capacitor in Fig. 1 acts almost as an open circuit." Indeed, if the capacitor in Fig. 1 were replaced by an open circuit, the resistor current would be zero, and thus the voltage across the resistor would also be zero; the output voltage would then be equal to the input voltage. At high frequencies, on the other hand, the capacitor voltage is small, as shown in Fig. 3(b), and if the frequency is made very high the capacitor voltage is almost zero; because of this, we say that "at very high frequencies, the capacitor in Fig. 1 acts almost as a short circuit." Indeed, if the capacitor in Fig. 1 were replaced by a short circuit, the output voltage would be zero.

The use of sinusoidal test signals

Since the sinusoidal signal occurs often in nature, it is a natural choice as a test signal for a variety of circuit studies. Also, as you will learn in theory classes, it turns out that if the behavior of a linear circuit (one consisting of linear elements) is known for sinusoidal input signals, its behavior for other types of input signals can be predicted as well. For these reasons, it is standard practice to investigate the behavior of circuits with sinusoidal inputs, both in theory and in the laboratory. We will adopt this practice in this lab.

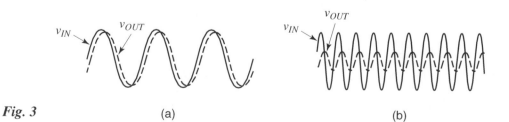

Fig. 3   (a)   (b)

Throughout our discussion we assume that the circuits we consider have been driven by a sinusoidal input for some time; any transients caused by the initial application of the input signal are supposed to have died out, so that the output (as well as all other voltages and currents in the circuit) is sinusoidal. The circuit is then said to be in the *sinusoidal steady state*.

Voltage gain

Consider a linear circuit and assume that, at a given frequency, the input and output signals are as shown in Fig. 4. The amplitude of the input is A_{IN}, and that of the output is A_{OUT}. To indicate to what extent the input signal has passed to the output, we define the voltage gain magnitude, G:

$$G = \frac{A_{OUT}}{A_{IN}} \tag{1}$$

The value of G depends on the circuit and the input frequency. For an amplifier, it can be much larger than 1. For the circuit of Fig. 1, it cannot exceed 1.

Output-Input phase shift

In Fig. 1, because the RC circuit shows some "inertia" (it takes time for the capacitor to charge or discharge before the output can rise or fall significantly), the various features of the sinusoidal output waveform (peaks, valleys, and zero crossings) appear to occur somewhat later than the corresponding features of the input waveform. We say that the output signal is *phase-shifted* with respect to the input signal. The phase difference between the output and the input can be determined with the help of Fig. 4. With the period T corresponding to 360 degrees, the phase difference between the two waveforms will be $[(\Delta t)/T] \times 360$ degrees, where $\Delta t = t_2 - t_1$, as shown in the figure. This is the phase by which the top waveform leads (is ahead of) the bottom one or, identically, the phase by which the bottom waveform lags (is behind) the top one. It is common to quote instead the phase by which the bottom waveform *leads* the top one; this is the *negative* of the above quantity and will be denoted by ϕ. Thus,

$$\varphi = -\frac{\Delta t}{T} \times 360° \tag{2}$$

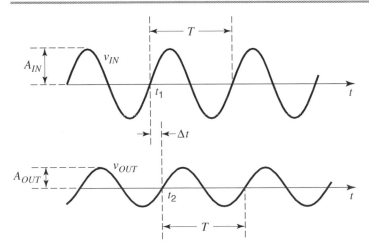

Fig. 4

Cutoff frequency

The values of G and ϕ depend on how fast the signal is varying in relation to the circuit's "speed." Consider, for example, the circuit of Fig. 1. We have seen in Experiment 6 that a measure of the circuit's "slowness" is its time constant, $\tau = RC$. Thus, a measure of the circuit's speed is the inverse, $1/\tau = 1/(RC)$. It is common to consider how the circuit behaves when the input's angular frequency ω (in rad/s) is equal to this measure. Since $\omega = 2\pi f$, where f is the frequency in Hz, we are interested in the behavior of the circuit at a critical frequency f_c for which $2\pi f_c = 1/(RC)$; thus f_c is given by

$$f_c = \frac{1}{2\pi RC} \qquad (3)$$

It is shown in theory classes that, when $f = f_c$, the value of G is $1/(\sqrt{2}) \approx 0.707$. The frequency f_c is called the *cutoff frequency*. The value of ϕ at this frequency can be shown to be -45 degrees.

Frequency response

The values of G and ϕ can be measured, or calculated, for various input frequencies f. The behavior of G versus f, together with that of ϕ versus f, taken together is called the *frequency response*. Sometimes, this term is used to refer to just one plot, that of G versus f. It is common to use a logarithmic frequency axis for such plots so that the behavior over a large range of frequencies can be conveniently displayed.

High-pass filter

Consider now the circuit of Fig. 5. Here the output is taken *across the resistor.* As before, at very low frequencies the capacitor will "act almost as an open circuit," which will virtually disconnect the output from the input. The output voltage can thus be expected to be very small at very low frequencies. At high frequencies, on the other hand, the capacitor will act almost as a short circuit, i.e. almost as if it were replaced by a direct connection from input to output, in which case the output voltage would be equal to the input voltage. Thus, for the circuit of Fig. 5, high-frequency input signals pass to the output, whereas low-frequency signals are rejected; for this reason, this circuit is said to be a *high-pass* filter. The voltage gain magnitude G can be defined again as above. When the frequency of the input is equal to f_c, where $f_c = 1/(2\pi RC)$, G takes the value of $1/(\sqrt{2}) \approx 0.707$, just as in the case of the low-pass filter. The quantity f_c is again called the cutoff frequency of the filter. At this frequency, the output leads the input by $+45$ degrees.

We are finally ready for the experiment.

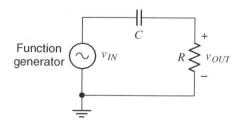

Fig. 5

A LOW-PASS FILTER AND ITS FREQUENCY RESPONSE

1. Connect the circuit of Fig. 1, using $R = 16$ kΩ and $C = 10$ nF. Set the input frequency at 1 kHz and its amplitude (peak value) at 2V. Monitor the input waveform on the scope's channel 1 and the output waveform on channel 2, with both channel inputs set at AC. Use $\times 1$ probes. The sensitivity of the two channels (V/div) should be the same, and the variable sensitivity control for each should be in the calibrated position. The trigger source selector should be at channel 1, the trigger coupling at AC, and the trigger slope at positive. You should be able to obtain a display like that in Fig. 4. Carefully establish the 0 level for each waveform by using the ground setting for each scope input, and align this level with a main graticule horizontal line, using the vertical position control.

2. Vary the frequency of the input signal over a wide frequency range (but keep the input amplitude fixed), and observe qualitatively the amplitude of the output.[1] Does what you see justify the name "low-pass filter"? Why?

3. Find the frequency at which the gain magnitude, G, drops to $1/(\sqrt{2}) \approx 0.707$. According to the background given above, this frequency should be the cutoff frequency, f_c. Record this value. Does it agree with the theoretical result, $f_c = 1/(2\pi RC)$?

4. Record the values of A_{IN} and A_{OUT} for input frequencies of 0.01, 0.1, 0.3, 1, 3, 10, and 100 kHz (all values are in kHz). Be sure that A_{IN} is maintained at 2 V at all frequencies; if necessary, adjust the amplitude control on your function generator.

5. Plot the gain magnitude, G (see Background above), versus frequency, using a *logarithmic* frequency axis; for convenience, you may want to trace Fig. 6 for this purpose.[2]

6. The phase shift between the output and the input, ϕ, was defined under Background above. Measure this quantity, with the input frequency equal to the cutoff frequency f_c that you determined in step 3. To make this measurement easier, move one of the two waveforms vertically, so that its 0 level coincides with the 0 level of the other waveform (use the ground input mode on the scope to establish the 0 levels accurately). Then Δt can be easily estimated as the horizontal displacement between two zero crossings, as shown in Fig. 4.[3] Is the phase shift at f_c approximately equal to the theoretically predicted value of -45 degrees?

7. Measure the phase shift at the frequencies you used in step 4. Make sure that the zero levels of the two waveforms coincide, as explained in the previous step. For high frequencies, where the output amplitude is small, you may need to change the sensitivity

---

[1] On analog scopes you may want to use the CHOP mode for slow sweep rates (e.g., 5 or more ms/division), and the ALT mode for faster sweep rates.

[2] G is often plotted in decibels (Appendix C); you are not asked to do so here.

[3] If there is a horizontal position control common to both channels on your scope, using it may help you read this displacement more easily.

Fig. 6

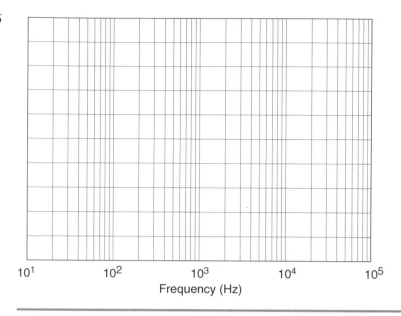

Frequency (Hz)

(V/div) of channel 2; every time you do so, you should be sure that the zero levels still coincide.

8. Plot the phase shift versus frequency, using a logarithmic frequency axis. You can trace Fig. 6 for this purpose.

A HIGH-PASS FILTER AND ITS FREQUENCY RESPONSE

9. Now build the high-pass filter of Fig. 5. Use $R = 16\,\text{k}\Omega$ and $C = 10\,\text{nF}$ as before. Vary the frequency of the input signal over a wide range. Observe qualitatively the amplitude of the output. Is your observation what you expected, and does it justify the name "high-pass filter"?

10. Perform appropriate measurements on this circuit so that you can plot G (see Background above) versus frequency. Use the same frequency values as in step 4. The horizontal axis for this plot should be logarithmic. Again, you may want to trace Fig. 6 for your plot.

11. Consider the frequency at which the value of G is equal to $1/\sqrt{2} \approx 0.707$. This is the cutoff frequency of the filter. Theory shows that this frequency should again be given by $f_c = 1/(2\pi RC)$, just as for the low-pass filter. Does your experimental result agree with this value?

TONE CONTROL

You will now pass audio signals through the filters you investigated above, and will listen to those signals. You need to use a CD or cassette player as your signal source.

Fig. 7

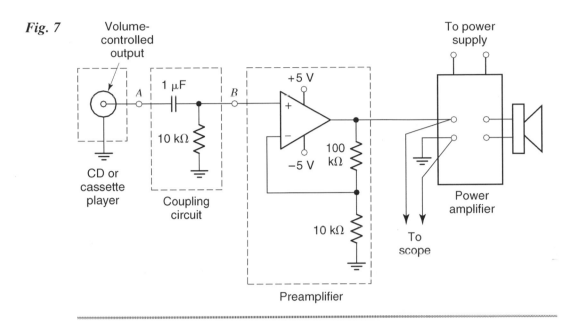

If no such player is provided, you may bring a portable one of your own; in that case, though, **operate it with batteries, to ensure safety when you make connections to it.** Although a stereo player provides two channels at its output (left and right), we will be using only one of them for simplicity.

12. Consider the circuit shown in Fig. 7, but do not build it yet. The circuit contains a preamplifier to increase the signal to a level suitable for driving the power amplifier. The 1 μF input capacitor in the coupling circuit acts almost like a short circuit at the frequencies of interest; it couples the AC signal from A to B, while preventing any accidental DC voltage from being coupled from your circuit to the player, or vice versa. The 10 kΩ resistor is used to allow the passage of a minute bias current needed by the input circuitry of the op amp (you do not need to concern yourself with this current; you can still make the assumption that the op amp's input current is almost zero).

The output on your player is assumed to have a volume control. (Such an output, if there is one, will most likely be the headphone output. Although this output is not optimum for feeding a high-quality audio amplifier, it can be conveniently used here since its volume control comes in handy.) If no volume control is provided on your player, replace the circuit between A and B with the circuit shown in Fig. 8.

Now build the circuit, keeping the volume control all the way down.

13. **Keep the speaker away from your ears and from those of others.** Make sure the volume control is all the way down. Turn on the power to the amplifier. Use your own favorite CD or one provided in your lab.[4] Set the player to PLAY, and increase the volume until you can hear the music at a comfortable level. The sound should be clean,

---

[4] For the purposes of this experiment, it is better to choose music with only a few acoustic instruments, including a bass guitar and cymbals.

Fig. 8

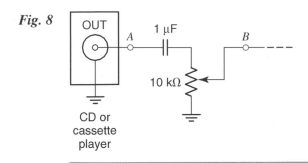

with no distortion. If this is not the case, lower your volume; if the problem persists, recheck your connections or ask your instructor for help. Observe the music waveform with the scope connected to the input of the power amplifier. Use a sweep rate of 1 ms/div.

14. Shut off the player and turn off all power. Insert your low-pass filter as shown in Fig. 9. The jumper wire is simply a wire that you can manually move from point D (solid line) to point E (broken line), or vice versa, for comparison purposes. A single-pole, double-throw switch could also be used. With the jumper at E, the signal goes through the filter before being fed to the power amplifier, and it will thus be modified by the filter's frequency response. With the jumper at D, the filter is being bypassed, and the power amplifier is driven by the op amp's output voltage, just as in step 13; thus, in this case, the system should sound as it does in step 13.

In steps 1-8, you studied how the low-pass filter affects a single-frequency input applied to its input. The music signal can be thought of as being composed of *many signals at various frequencies*. The filter is a linear circuit and will treat each of these signals as if it were present by itself, in the manner studied above.

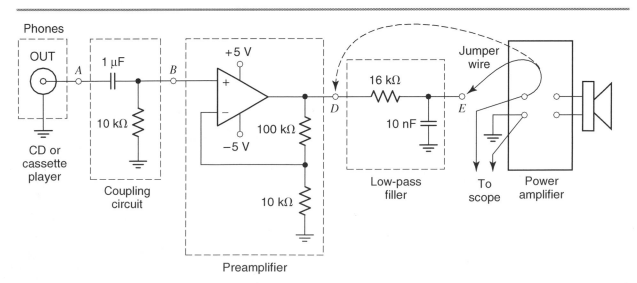

Fig. 9

Fig. 10

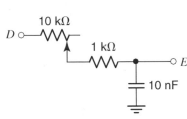

Fig. 11

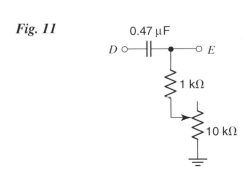

15. With the jumper at D, listen to the music again. The sound should be as before (if it is not, debug your circuit). Listen for a minute or so, and pay special attention to the very low frequencies in the sound (e.g., those from a bass guitar) and the very high frequencies (e.g., those from a cymbal).

16. Without turning anything off, move the jumper from D to E. Listen carefully. What do you observe? What can you say about the presence or absence of bass (low) frequencies and of treble (high frequencies) in the sound? Which instruments are favored by the filter? Relate what you hear to the term "low-pass filter." Be sure that you hear the difference the filter makes by moving the jumper back and forth between D and E.

17. Look at the music waveform, with the scope connected to the input of the power amplifier. Use a sweep rate of 1 ms/div. Again move the jumper back and forth between D and E. What do you observe? Can you relate the difference you see to the difference you hear?

18. Roughly speaking, when the jumper is at E, the filter allows signals at frequencies below the cutoff frequency you determined above to pass to the amplifier. You can make this frequency variable by replacing the fixed resistor in the low-pass filter by a variable one. One way to do this is shown in Fig. 10. Be sure that one of the two terminals you use for the potentiometer is the slider. The resistor in series with the potentiometer is used to prevent the total resistance from becoming zero when the potentiometer slider is moved all the way to the left, in which case the op amp may have difficulty driving the capacitor directly. *Before building this circuit*, answer this question: Over what range can the cutoff frequency of the filter be varied as you turn the potentiometer from one of its extremes to the other?

19. Replace the low-pass filter in Fig. 9 by the one in Fig. 10. Listen again to the music, and observe it with the scope. Vary the potentiometer, and notice the effect of different cutoff frequencies on the music. How does the potentiometer affect the high frequencies (treble) in the sound? You have just made a simple *treble cut control*.

20. Now replace the low-pass filter by the high-pass filter in Fig. 11; the resistor of Fig. 5 has been made variable, as in the previous step. Again, vary the potentiometer and observe the effect of different cutoff frequencies on the sound you hear, as well as on the waveform on the scope display (you may need to use a different sweep rate, say 10 ms/div). How are the low frequencies (bass) affected? This is a simple *bass cut control*.

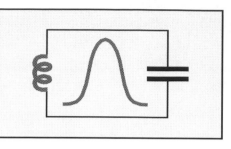

EXPERIMENT 8

LC CIRCUITS, RESONANCE, AND TRANSFORMERS

Objective

In this experiment you will work with circuits that contain capacitors and inductors. You will observe resonance and the related fact that such circuits respond better to signals of certain frequencies (a property called selectivity). You will use the magnetic core of the inductor as an antenna for the magnetic component of radio waves. Throughout this experiment, a transformer will be used for signal coupling.

Preparation

You may want to study LC circuits and transformers in your circuits text. However, even if you don't, the background provided below should be adequate for you to do the experiment.

NOTE: Throughout this experiment, *try to use wires as short as possible* unless explicitly stated otherwise. The circuits you will build have to be tested at radio frequencies. Long wires can act as antennas and interfere with your measurements.

RESONANT CIRCUITS

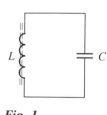

Fig. 1

1. Consider the resonant circuit consisting of an inductor and a capacitor, as shown in Fig. 1. This circuit is also referred to as an LC tank. You may have studied this circuit in physics, and it is discussed in detail in circuits classes. The inductor consists of many turns of insulated wire and exhibits an inductance that depends on the inductor's geometry, the number of turns, and the material used for its core (around which the wire is wound). The inductor in this experiment uses a ferrite material of high magnetic permeability as its core. This material is indicated schematically by the broken lines in Fig. 1. When a capacitor is connected across the inductor, the resulting LC resonant circuit has a preferred frequency, called *resonance frequency*. This is the frequency at which its voltage and current oscillate if the circuit is kicked by an electrical

Fig. 2

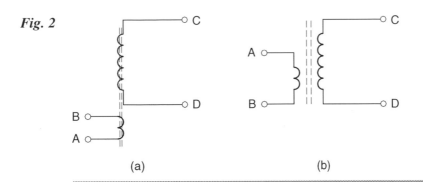

(a)   (b)

impulse and then let free. In physics and circuits classes it is proven that the resonance frequency is given by

$$f_{resonance} = \frac{1}{2\pi\sqrt{LC}} \qquad (1)$$

where L is the inductance and C the capacitance. If excited by an external signal, the circuit's voltage and current respond to that signal strongly if its frequency is at or near the circuit's resonance frequency (think of a mechanical analog, involving a pendulum, to which you give pushes at the rate it "wants" to swing).

TRANSFORMERS

2.   One way to couple energy to the inductor is through a second winding, as shown in Fig. 2(a). If a time-varying current is made to flow in this winding, it will induce a magnetic field. This field will cut across the other winding and will induce a voltage at its ends. We thus have a *transformer*. The magnetic coupling is made tight by using the same ferrite core for both inductors. Although the drawing in Fig. 2(a) corresponds to the physical construction of the transformer we will be using in this experiment, from now on we will be using instead the widely adopted symbol shown in Fig. 2(b).

   Check your transformer, and code it as follows: C should indicate the terminal of the larger winding, *which is further away from the small winding*; the other terminal of the larger winding should be D. The terminal of the small winding closest to terminal D of the large winding should be B; finally, the other terminal of the small winding should be A. It is important to keep the relative position of the various terminals in mind; among other things, terminal C is further away from the small winding than terminal D is, and this means that it has a smaller parasitic capacitance[1] to the small winding. This will turn out to be important for the correct operation of some of the circuits we will be discussing.

---

[1] "Parasitic capacitance" in our case means an unintended capacitance that exists because of the close proximity of two conductors. Often such a capacitance can adversely affect the operation of a circuit.

Fig. 3

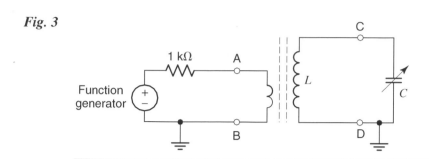

FREQUENCY RESPONSE OF A RESONANT CIRCUIT; BAND-PASS FILTERING

3. To study LC circuit resonance, set up the circuit of Fig. 3. The outer conductor of the function generator's output connector is ground; the inner conductor should be connected to the 1 kΩ resistor. The arrow through the capacitor indicates that this capacitor is variable (for example, from 10 pF to 150 pF); set this capacitor near the middle of its range. To monitor the voltage across the LC tank, use the scope. The input capacitance of the probe will then appear in parallel with C, and if it is too large (as it is in the case of a × 1 probe) it will seriously interfere with your measurement. To make this effect less serious, use a × 10 probe, which has a much smaller input capacitance (e.g., 20 pF). The right ground in Fig. 3 denotes the ground of the scope, which is connected to the clip of the probe; the inner conductor of the probe should be connected to the upper terminal of the capacitor. Use a 1 V amplitude for the output of the signal generator. Vary the signal frequency from about 0.5 MHz to 2 MHz, and observe the amplitude of the voltage across the LC tank. At some frequency, the amplitude of the voltage should be the largest. We then say that the circuit exhibits resonance. The effect of resonance should be dramatic for this circuit. What is the resonance frequency?

4. Plot the amplitude of the voltage across the capacitor versus the frequency of the input signal, with the amplitude of the latter at 1 V. Be sure that you include frequencies both below and above the resonance frequency, and use a sufficient number of points near the resonance frequency. Note the shape of this plot. We say that the circuit exhibits *selectivity* because the amplitude of its output can be large or small, depending on whether the input frequency is near the resonance frequency or not. In other words, there is a range (or band) of frequencies at which the input causes a large output; the output becomes small as the input frequency is moved outside that range. For this reason, this circuit is a type of *band-pass* filter.[2]

---

[2]A measurement of the generator signal versus frequency would show that the amplitude of that signal is almost independent of frequency. Thus the frequency response magnitude (Eq. 1 in Experiment 7) of this circuit is approximately the same as the curve you plotted above.

Fig. 5

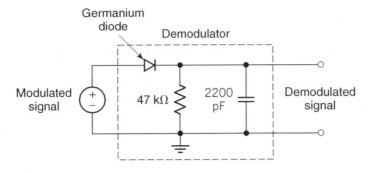

8. Hook up the circuit of Fig. 5, and observe its output waveform on channel 1 of the scope. Use the DC input coupling for the latter. Leave the audio signal connected to channel 2, with triggering from channel 2, as before.

9. Try various other capacitance values in the circuit of Fig. 5, and explain qualitatively what you see. What happens if the capacitance is (a) too small or (b) too large? Why? What is an appropriate capacitance value for recovering the original audio shape?

    When finished, do not disassemble your circuit; you will need it later.

LISTENING TO THE DEMODULATED SIGNAL

10. Lower the RF amplitude so that the peak-to-peak, demodulated signal is only a few tens of mV; this amplitude is typical of medium-strength signals in the input stage of AM receivers.

11. To amplify the weak demodulated signal, construct the voltage amplifier of Fig. 3. How much voltage gain do you expect out of this circuit when the potentiometer is set all the way up?

12. Connect the output of your demodulator of Fig. 5 to the input of the preamplifier of Fig. 3. Check the output of the volume control with the scope to make sure that the whole circuit works properly. You should be able to observe the amplified demodulated signal, which should look approximately like your original audio signal. When finished, turn the volume control all the way down.

13. Connect the output of the volume control to the power amplifier and the output of the latter to the speaker, as shown in Fig. 6. Do not forget to connect a power source to the power amplifier (see Experiments 3 and 5). **Keep the loudspeaker away from your ears and from the ears of your colleagues.** Turn up the volume control slowly. You should be able to hear the demodulated audio signal clearly. Keep the volume sufficiently low so that you do not disturb other people in the lab. When finished, turn the function generators off.

Fig. 6

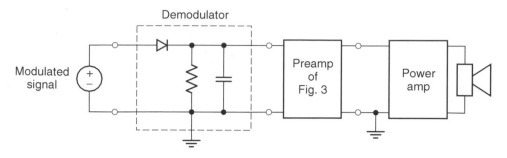

14.   You have just built a receiver that can receive a modulated RF signal and recover the audio information in it. The trouble is, this receiver is *not* selective; it will pick up all strong signals at its input, demodulate them, and play them simultaneously. To see this, disconnect the function generator from the input of the demodulator and, instead, connect there one end of a long wire (e.g., several meters long). Extend this wire, and leave its other end unconnected. This wire will act as an antenna. If several stations have a strong enough signal where you are, you should hear several programs, all mixed together. To avoid this problem, you will need to add some means for providing *selectivity* to your receiver. You can do this by using a resonant circuit, as described below.

Before proceeding, turn off all power supplies and remove the antenna.

RADIO RECEIVER

15.   Build the radio receiver circuit shown in Fig. 7, but do not turn it on yet. The *tuner* contains a resonant circuit that consists of an inductor and a capacitor, with its output obtained through transformer coupling. This circuit has been studied in Experiment 8; see that experiment for the meaning of terminals A, B, C, D in the transformer (important). The inductor and variable capacitor used should be such that the resonance frequency can be varied approximately in the AM band (530–1700 kHz; see also Experiment 8). This circuit selects the signal of the desired station and feeds it to the demodulator circuit containing a germanium diode (see Background above). The output of this circuit is fed to the amplifier of Fig. 3 and provides a large enough signal to drive the power amplifier. *Be sure that you use wires as short as possible. Be careful with the various ground connections. Turn the volume control all the way down.*

16.   You are now ready to pick up radio stations. **Keep the loudspeaker away from your ears and from the ears of your colleagues.** Turn on the power to both preamplifier and amplifier. Turn up the volume control slowly. Try tuning your receiver by turning the variable capacitor knob. If there are strong stations in the vicinity, and if the building construction is such that it allows the waves to enter the room, reception will be easy. Reception may be improved by changing the orientation of the ferrite transformer in the tuner (Experiment 8). If the signals are too weak to receive in this way, try connecting a wire antenna (a wire a few meters long) to the upper end of one of the two windings of the transformer. You may need to retune after making this connection. Reception may be better near a window. If you are still encountering difficulties, your reception may be improved as explained in the following two steps.

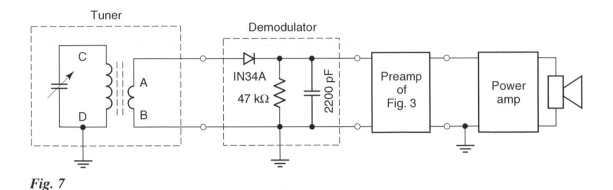

Fig. 7

17. If you cannot receive a sufficient number of stations, the problem may be that your receiver cannot tune the entire AM band. If your tuning range does not include the upper portion of the AM band, you can decrease the inductance a little by sliding the ferrite core partially out, *in the direction corresponding to the downward direction* in Fig. 7, as explained in Experiment 8, and retuning the capacitor. If you still cannot hear any station, reception may be too weak at your location. Try increasing the amplifier's gain by increasing the value of R_2 in Fig. 3 (e.g., to 150 kΩ).

18. If you hear interference, it may be coming from the many instruments operating in the room. Making all connecting wires as short as possible should help keep interference low; as we have repeatedly pointed out, long wires can act as antennas for interference. Try turning off as many of the instruments as possible, except, of course, for the supplies that provide power to your amplifiers. Unfortunately, such supplies may contain switching circuits, which can cause high-frequency interference. If you suspect that this is the case, you can try to prevent this interference from reaching your receiver by connecting capacitors (in the several 1000 pF range) across the power supplies, either at the supply output or where they are connected to your amplifiers. The high-frequency currents will then find it easier to go through these capacitors rather than through your circuit; for this reason, capacitors used in this way are sometimes referred to as *bypass* or *decoupling* capacitors.[2] If you still cannot get rid of the interference or if you cannot pick up any stations, ask your lab instructor for help.

19. Using $\times 10$ probes, observe the signals at the output of the tuner, the demodulator, the preamplifier, and the amplifier. Make sure you have come full circle and that you understand everything you see.

---

[2]More on supply bypassing can be found in Appendix D.

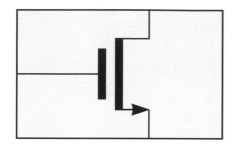

EXPERIMENT 11
MOSFET CHARACTERISTICS AND APPLICATIONS

Objective   In this experiment you will study the *i-v* characteristics of an MOS transistor. You will use the MOSFET as a variable resistor and as a switch.

BACKGROUND

The MOS (metal-oxide-semiconductor) transistor (or MOSFET) is the basic building block of most computer chips, as well as of chips that include analog and digital circuits. In this lab, we will work with what is called an n-channel MOS transistor. The internal structure and operation of this device, as well as of its complement, the p-channel MOSFET, are studied in semiconductor device courses. Here we will concern ourselves only with external *i-v* behavior.

A common symbol for the n-channel MOS transistor is shown in Fig. 1(a). Of the terminals shown, the ones we will focus on are the *source*, the *drain*, and the *gate*. In the transistor we will be working with in this lab, the fourth terminal, labeled *body* in Fig. 1(a), is not accessible externally; rather, it is internally connected to the source, as shown by the broken line. For simplicity, then, we will use the symbol shown in Fig. 1(b).

Fig. 1   (a)   (b)

Fig. 2

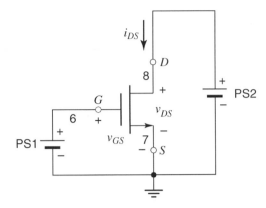

Fig. 3

CD4007

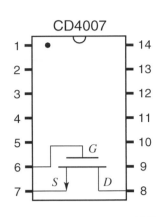

To study the MOS transistor, we can connect two external voltage sources to it, as shown in Fig. 2. These provide the drain-source voltage v_{DS} and the gate-source voltage v_{GS}. The voltage v_{DS} may cause a drain-to-source current i_{DS} as shown, provided that there is a path for this current from the drain (through the device) to the source. Whether such a path exists or not depends on the value of v_{GS}. For small values of v_{GS}, no such path is established within the device, and i_{DS} is zero. The device then looks like an open circuit between the drain and the source. For a sufficiently high v_{GS}, an internal current path, called a *channel*, is established between the drain and the source.[1] Now a current i_{DS} can flow. The precise value of v_{GS} determines how easy it is for the channel to conduct the current; the higher the v_{GS} value, the easier such conduction is and the larger the value of i_{DS}, other things being equal. For a given v_{GS}, the value of i_{DS} will also depend on v_{DS} and will tend to increase with the latter.

It should be noted that the only DC current in the device is the drain-to-source current i_{DS}. The gate is internally separated by an insulator from the channel, so the gate current is practically zero.

MOSFET I-V CHARACTERISTICS

1. Hook up the circuit of Fig. 2. This circuit will be used in the following steps to investigate the *i-v* characteristics of the n-channel MOSFET. The chip used in this experiment is a CD4007, containing six MOSFETs. We will use only one of them, as shown in the pin assignment in Fig. 3.

2. Set $v_{GS} = 5$ V. Measure the drain current, i_{DS}, versus the drain-source voltage, v_{DS}, from 0 to 5 V. Make sure you take measurements at a sufficient number of v_{DS} values since you will later need to plot i_{DS} versus v_{DS}. Include a point at $v_{DS} = 0.1$ V for later use.

3. Repeat the entire step 2 for $v_{GS} = 3$ V and $v_{DS} = 1$ V.

---

[1] This channel consists of mobile electrons, which are negatively charged—hence the name n-channel MOSFET.

4. With $v_{DS} = 5$ V, determine the value of v_{GS} at which the current i_{DS} becomes negligible; assume that for our purposes this means $5\,\mu$A. This value of v_{GS} is close to the so-called *threshold voltage* of the transistor, and it is positive for an "enhancement-mode" MOSFET, which is what we are working with here.[2]

5. Using the data you have collected in steps 2 and 3, plot a family of curves for the drain current, i_{DS}, versus the drain-source voltage, v_{DS}, from 0 to 5 V, with v_{GS} as a parameter. Use a *single* set of v_{DS} - i_{DS} axes for this plot. There should be one curve for each v_{GS} value (1 V, 3 V, and 5 V) on this family of curves. Label each curve with the corresponding v_{GS} value.

6. You should be able to observe on the above plot that, for each curve, the current tends to a constant (or, as we say, saturates) as v_{DS} is made large. What is, approximately, the saturation value of the current for each of the three v_{GS} values?

THE MOSFET AS A VOLTAGE-CONTROLLED RESISTOR

7. Verify that the curves obtained in step 5 pass through the origin and that their shape is nearly a straight line for *sufficiently small* v_{DS} values. Thus, in that region i_{DS} is approximately proportional to v_{DS}, and Ohm's law is approximately satisfied; that is, we have a nearly linear resistor. The resistance $R = v_{DS}/i_{DS}$ is given by the inverse of the slope of the curves near $v_{DS} = 0$. Since the slope depends on v_{GS}, the latter can be used to control the resistance value. Determine this resistance *graphically* for v_{GS} values of 5 V, 3 V, and 1 V.

8. Form the circuit of Fig. 4(a), by *disconnecting* PS2 and the DMMs from the circuit of Fig. 2. Connect an ohmmeter between the drain and the source, and verify that the resistance across these two terminals can be varied by varying v_{GS}. Verify the values

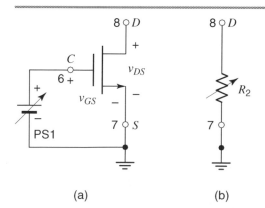

Fig. 4   (a)   (b)

---

[2]Another variety is the depletion-mode MOSFET, for which the threshold voltage is negative. We will not be using depletion-mode MOSFETs in this lab.

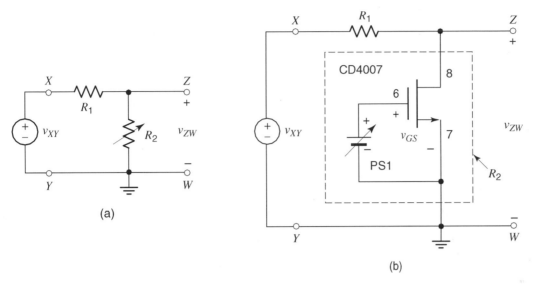

Fig. 5

you calculated in step 7. Also, determine the resistance when $v_{GS} = 0$. When finished with this part, *disconnect the ohmmeter from the circuit*.

From the results so far, it is evident that the circuit of Fig. 4(a) is a *voltage-controlled resistor*; that is, it is equivalent to Fig. 4(b), where the value of R_2 depends on v_{GS}. The arrow through the resistor indicates that its resistance can be varied. (Keep in mind that this resistor is linear provided that the drain-source voltage is kept sufficiently small; otherwise, its current will not be proportional to the voltage across it.)

9. You can use the fact that the MOSFET can be used as a voltage-controlled resistor, to make a voltage-controlled potentiometer. Consider the voltage divider shown in Fig. 5(a). R_1 is a conventional, fixed resistor, whereas R_2 is a variable resistor. If a voltage v_{XY} is applied to the circuit as shown, the output voltage v_{ZW} will be

$$v_{ZW} = \frac{R_2}{R_1 + R_2} v_{XY} \tag{1}$$

In the circuit of Fig. 5(a), R_2 can be replaced by the MOSFET, as suggested by the equivalence of the circuits in Figs. 4(a) and 4(b). Thus, the circuit of Fig. 5(b) results (*do not build this circuit yet*). Now, one can vary the division ratio $R_2/(R_1 + R_2)$ by varying R_2 through v_{GS}.

Based on your results from steps 7 and 8, calculate a value to be used for R_1 so that the voltage division ratio, $R_2/(R_1 + R_2)$, can be varied from the value of 1 to a value of about 0.1, as v_{GS} is varied from 0 to 5 V.

10. Now build the circuit of Fig. 5(b). Keep in mind that the MOSFET behaves as a linear resistor *only* if the voltage between its drain and source is sufficiently small; thus, *avoid* large v_{ZW} values. Do *not* use an ohmmeter in this step. Verify that the voltage division ratio, $v_{ZW}/v_{XY} = R_2/(R_1 + R_2)$, can be varied in the range expected from step 9 by varying v_{GS}.

THE MOSFET AS A SWITCH

11. From the results obtained so far, you can see that if v_{GS} is sufficiently small, the MOSFET behaves as an open circuit between the drain and the source; thus, the drain-source path in Fig. 4 acts as an open switch in this case. Also, you can see that with v_{GS} large, the MOSFET presents a rather small resistance between the source and the drain (always assuming that the drain-source voltage is small). If that resistance were zero, the MOSFET would behave as a closed ideal switch in this case; since the resistance is not zero, we can say that it behaves as a closed nonideal switch (essentially, it behaves as a closed ideal switch with some resistance in series with it).

Thus, the MOSFET can be viewed as a voltage-controlled switch; the switch closes if v_{GS} is made large and opens if v_{GS} is made small.

12. The resistance of the closed MOSFET switch above is significant because the MOSFETs on the chip used in the above steps are not meant to operate as switches per se. There are other chips which are explicitly designed for this purpose. They contain transistors designed to have a small resistance when the magnitude of their v_{GS} is large.[3] Such chips also contain internal circuitry that develops the proper v_{GS} values for turning the switch on or off, depending on the value of an external control voltage supplied by the user. You will now use such a chip.

13. The chip we will be working with is the CD4066 or equivalent type, which contains four switches. The pin assignment for the switch we will be using is shown in Fig. 6(a). Connect the chip as shown, *but do not turn on the power yet*. The switch will close if an externally applied control voltage $v_{CONTROL}$ is made high, and it will open if the control voltage is made low (*do not test this yet*). This behavior is indicated schematically in Fig. 6(b); in this figure, the ±5 V power supply connections in Fig. 6(a) are not shown, but keep in mind that they are *required* for the switch to operate.

14. Verify the behavior described in the previous step by using the connection indicated in Fig. 7. Ground terminal K, and measure the resistance between terminal L and ground with an ohmmeter. Use a control voltage $V_{CONTROL}$ of 2 V to close the switch, and −2 V to open it. What are the corresponding resistance values? Can they reasonably be thought to correspond to a short and an open circuit, respectively?

A CHOPPER

15. The circuit in Fig. 8(a) is called a *chopper*, and is used in communications and instrumentation. When the switch is closed, the output is connected to the input and v_{OUT} is

---

[3] To maintain this low resistance even when their terminal voltages vary considerably, two MOSFETs inside such chips (one n-channel and one p-channel) are connected in parallel to form a switch. You do not need to concern yourself with this at this time.

Fig. 6

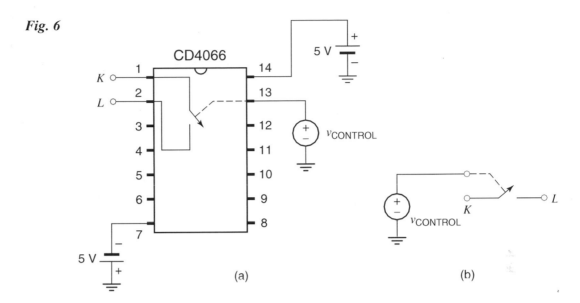

(a)   (b)

Fig. 7

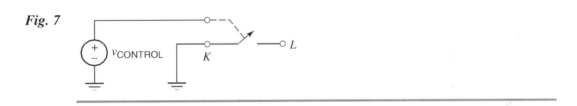

equal to v_{IN}; when the switch is open, the output is disconnected from the input and v_{OUT} equals zero. This operation is illustrated in Fig. 8(b).

Assume that the input voltage is a 1 kHz sinusoidal voltage with an amplitude of 1 V and that the control voltage is a 10 kHz square wave, taking values +2 V and − 2 V. *Without connecting the circuit*, sketch the expected waveform at the output, based on the behavior you observed in the previous step.

16. Now connect the circuit of Fig. 8(a), using the values given in the previous step. Be sure the power supplies are connected as has been shown in Fig. 6(a) and are turned on. You will need two function generators. Use the scope to adjust the generator that provides the control voltage, so that it is a 10 kHz square wave with values − 2 V and 2 V at its peaks and bottoms, respectively (if your generator does not provide such large values, smaller values can be tried). Then, use the scope's channel 1 to observe the input voltage (a 1 kHz sinusoidal voltage with a 1 V amplitude) and channel 2 to observe the output voltage. Trigger from channel 1. You will need to adjust *slowly and very carefully* the frequency of the control voltage generator to get a stable display. This adjustment may be tricky. Verify that the output waveform is as expected in step 15.

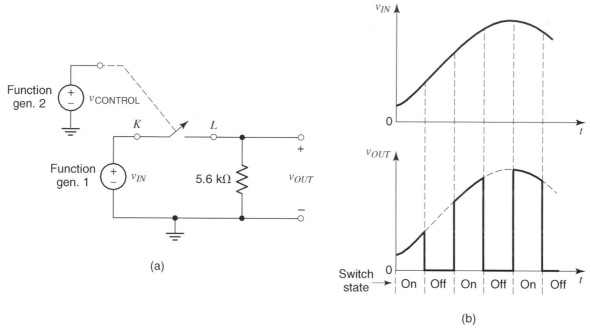

Fig. 8

A TRACK-AND-HOLD CIRCUIT

17. Consider the circuit of Fig. 9(a). It operates as follows:

(a) When the switch is closed, the output is equal to the input (i.e., the output tracks the input).

(b) When the switch is opened, the voltage on the capacitor cannot change anymore and remains at the value it had at the moment the switch opened.

This operation is illustrated in Fig. 9(b). As seen, when the switch is opened the capacitor holds the above value until the switch closes again, at which point the output again becomes equal to the input and tracks it. This is a *track-and-hold* circuit. It has many uses; for example, it is found at the input of analog-to-digital converters; during the intervals that the track-and-hold's output is held constant, the converter processes it to convert it to a digital code. This code can then be fed to a computer, to a digital communications link for transmission, or to a digital signal processor for further processing. Analog-to-digital converters are used, for example, in recording studios to convert music signals to digital words, which can then be stored on compact discs.

18. Assume that the input voltage is again a 1 kHz sinusoidal voltage with an amplitude of 1 V and that the control voltage is a 10 kHz square wave, taking values +2 V and −2 V. *Without building the track-and-hold circuit*, sketch the expected waveform at the output, based on the description of its operation in step 17.

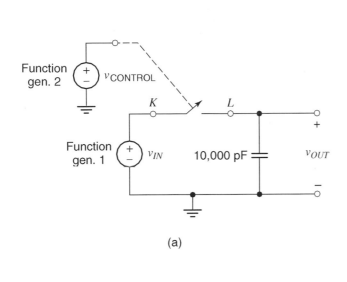

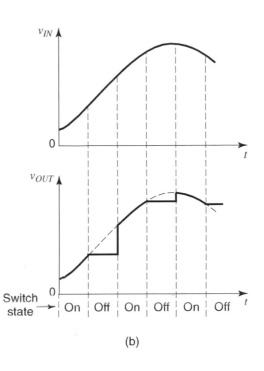

(a)

(b)

Fig. 9

19. Now build the circuit and try it. Use the scope as explained in step 16. Again, *careful fine-tuning of the control voltage generator's frequency will be needed obtain a stable display.* Is the output waveform as you expected in step 18?

NOTE 1: If your input signal generator has difficulty in driving this circuit, insert a resistance of a few hundred ohms between the signal generator and point *K*. This will ease the load on the signal generator.

NOTE 2: The output waveform of the track-and-hold contains a varying part and a constant part for each cycle of the control voltage. The time duration of the varying part can be made short by making the duty cycle of the control voltage small (a duty cycle is the percentage of the control voltage period during which the square wave has a positive value).

A SAMPLE-AND-HOLD CIRCUIT

20. If you still have time, you can ponder this question: Can you think of a way to further process the output signal of the circuit in Fig. 9(a), so that the varying part in each control voltage period is eliminated? You would need an additional track-and-hold circuit to do this. Do not build such a circuit; just explain how you would build it if you had to. Circuits that produce such waveforms at their output from a continuously varying input signal are called *sample-and-hold* circuits (or, more accurately, *sample-delay-hold circuits*). Sometimes, even the track-and-hold circuit is called a sample-and-hold circuit.

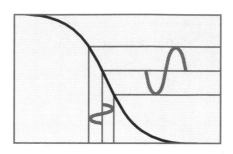

EXPERIMENT 12

PRINCIPLES OF AMPLIFICATION

Objective   In this experiment you will study the input-output characteristics of an amplifier, using an n-channel MOS transistor (MOSFET). You will bias the transistor at an appropriate point for use as an amplifier, and will determine its voltage gain. You will also study the effect of the bias point on distortion.

Background   If you haven't done Experiment 11, study its background section.

A MOSFET-RESISTOR INVERTER

1. Hook up the circuit of Fig. 1, using connections as short as possible, and with the voltage of PS1 set to zero for now. The chip that will be used in this experiment is the CD4007, the same chip used in the first half of Experiment 11. Fig. 3 of that experiment shows the pin assignment for the transistor we will be using. The circuit of Fig. 1 will be used in the following steps to investigate the use of a MOSFET as an amplifier.

2. The gate-source voltage of the MOSFET, v_{GS}, is equal to the voltage of power supply PS1. When this voltage is 0, the transistor current is virtually zero (this is the case for an enhancement-mode MOSFET, such as the one we will be using in this lab). If v_{GS} is gradually raised, initially the transistor current will remain at zero. When this voltage is raised sufficiently (to about the value of the threshold voltage of the transistor), the drain-to-source current i_{DS} of the MOSFET will rise above zero. This will cause a voltage drop $v_R = Ri_{DS}$ across the load resistor R. Using KVL you can see that the drain-source voltage v_{DS} will be equal to $V_{DD} - v_R$, and will thus be less than V_{DD}. As v_{GS} is increased, i_{DS} and thus v_R will also increase; therefore, v_{DS} will decrease. Eventually, v_{DS} can become so small that it limits further increases in i_{DS} and thus in v_R. Think what, qualitatively, a plot of v_{DS} versus v_{GS} will look like, without powering up the circuit.

3. Now take measurements and produce a plot of v_{DS} versus v_{GS}, with v_{GS} between 0 and 5 V. Make sure that you take enough measurements to adequately reproduce the steep part of the plot.

Fig. 1

Fig. 2

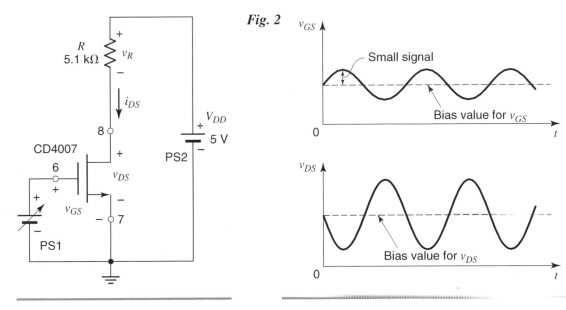

4. In the plot produced in the previous step, consider the point at which $v_{DS} = 2.5$ V. What is the corresponding value of v_{GS} as obtained from this plot? Record this value.

5. Your plot should be steep around the point defined in the previous step. If v_{GS} varies around this point, the corresponding variation of v_{DS} should be larger. What is, approximately, the value of the slope $(\Delta v_{DS})/(\Delta v_{GS})$ around this point?[1] Is it positive or negative? Why? The behavior you see is often said to be *inverting*. Why this name?

6. Set PS1 so that $v_{DS} = 2.5$ V. We will use this as the *bias* value of v_{DS} (the use of this quantity will become clear shortly). Measure the corresponding value of v_{GS}; this will be called the *bias* value of v_{GS}; it should agree with the value you found in step 4. Record the bias values of v_{GS} and v_{DS} for further use.

7. Increase v_{GS} by a small amount Δv_{GS}, say 0.1 V, above its bias value. Record the corresponding change Δv_{DS} of v_{DS}. From these results, calculate the *small-signal voltage gain* $(\Delta v_{DS})/(\Delta v_{GS})$. Does it agree with the value you found in step 5?

8. Repeat the above step, using instead a *decrease* of v_{GS} from its bias value (i.e., Δv_{GS} should now be -0.1 V).

9. Move the knob of PS1 back and forth slowly and continuously around the bias point, by about 0.1 V in either direction. Qualitatively observe the variation of v_{DS}.

10. What do you think v_{DS} would look like if Δv_{GS} were a sinusoidal voltage with an amplitude of 0.25 V?

---

[1] ΔV denotes a small change in V.

11. According to what you found so far, you can deduce that if you can superimpose a small signal to the gate-source bias voltage, so that it changes v_{GS} *around* the bias value, you will obtain an amplified version of this signal as a *variation* in v_{DS}. In other words, if v_{GS} looks like Fig. 2(a), v_{DS} will look like Fig. 2(b). The next step deals with a way to produce a voltage v_{GS} as in Fig. 2(a).

ADDING A SIGNAL TO A DC BIAS VOLTAGE

12. Consider the circuit of Fig. 3. It contains a DC source V_{DC} and an AC source $v_{AC}(t)$. The output voltage of the circuit is affected by both of these voltages. If the frequency of the signal is much higher than $1/(2\pi RC)$, the output of the circuit will almost be $v(t) = V_{DC} + v_{AC}(t)$, as indicated in the figure. To see this, consider the effect of each voltage source in the circuit by itself, as explained below.

Assume, first, that $v_{AC}(t) = 0$; this is equivalent to assuming that the signal generator is replaced by a short circuit, as shown in Fig. 4(a). Then the capacitor charges up to the DC voltage value, that is, $v_C = V_{DC}$. Using Kirchhoff's voltage law, we conclude that, in the steady state, $v = V_{DC}$, as indicated in the figure.

Assume now that the DC voltage in Fig. 3 is zero instead; this is equivalent to assuming that the DC source is replaced by a short, as shown in Fig. 4(b). The capacitor and resistor now form a high-pass filter with cutoff frequency $1/(2\pi RC)$ (see Experiment 7). If the frequency of the signal is much higher than this, the signal will pass to the output; in other words, $v(t) = v_{AC}(t)$, as indicated in the figure.

Since the circuit is *linear*, superposition will apply: When *both* voltage sources in Fig. 3 are active, the output voltage will be the sum of the individual results in Figs. 4(a) and (b), that is, $v(t) = V_{DC} + v_{AC}(t)$, as indicated in Fig. 3. Thus, we have a circuit in which the AC voltage is coupled to the output through the capacitor (which is, for this reason, called a *coupling capacitor*) and is added to the DC voltage. At the same time, the capacitor blocks the DC voltage from reaching the AC source.

Hook up the circuit of Fig. 3. (CAUTION: It is unlikely that the 100 nF capacitor will have polarity indicated on it, but if it does, make sure to observe it, as shown in Fig. 3.) Use a DC voltage of 2 or 3 volts and a signal voltage with an amplitude of about 0.25 V and a frequency of 1 kHz. [Is this frequency much higher than

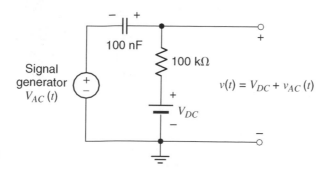

Fig. 3

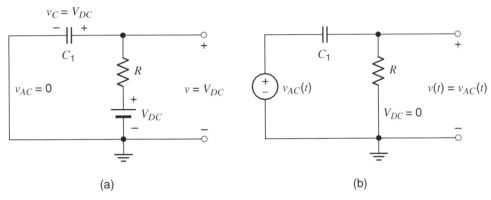

Fig. 4

1/($2\pi RC$) for this circuit?] Use the oscilloscope to observe the output signal. Make sure that you use the DC setting of the scope's input switch so that you can observe the *total* voltage and not just its AC variation. Do the following:

(a) Use the amplitude control on the function generator to reduce the generator's output to zero, and verify that $v = V_{DC}$.

(b) Use the variable voltage control on the power supply to reduce V_{DC} to zero, and raise the amplitude of the function generator output; verify that $v(t) = v_{AC}(t)$.

(c) Finally, allow both V_{DC} and $v_{AC}(t)$ to be nonzero, and verify that $v(t) = V_{DC} + v_{AC}(t)$.

OBTAINING AC AMPLIFICATION WITH A MOSFET

13. Remove PS1 from the circuit of Fig. 1, and instead use the circuit of Fig. 3 to produce the voltage v_{GS}, as shown in Fig. 5 (for now, set the signal amplitude to 0). The resistor shown connected across the output is meant to indicate any load the amplifier is supposed to drive; for our purposes, this load will be just a scope probe. The probe has a large input resistance (typically 1 MΩ) and can be connected there without disturbing the operation of the amplifier. Capacitor C_2 is a coupling capacitor, which allows the AC variation of v_{DS} to appear across the load but blocks the DC value of that voltage. Essentially, C_2 forms a high-pass filter together with the load resistance. This filter passes the (high-frequency) AC variation of v_{DS} to the load but rejects the DC (i.e., zero-frequency) part of v_{DS}. (In the unlikely event that the capacitors indicate polarity on them, make sure to observe it, as shown in Fig. 5.)

14. Keep the input signal amplitude at zero. Adjust the value of V_{DC} until v_{GS} and v_{DS} assume the bias values determined in step 6.

15. Now try the circuit as an amplifier, using a 1 kHz input signal with an amplitude of 0.25 V. Use the scope to study the waveforms of v_{IN}, v_{GS}, v_{DS}, and v_{OUT}, as well as how these are related to one another. You can use both channels of the scope for this

Fig. 5

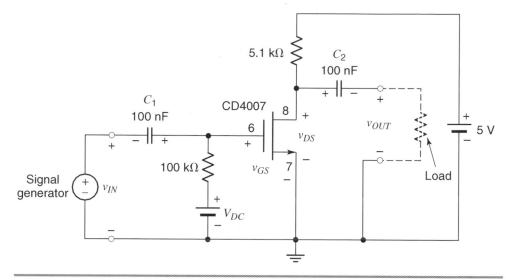

purpose. Use the center horizontal line as your zero reference (both traces should be positioned at that line when the input switches for both channels are at the ground position). A convenient vertical sensitivity to use for both channels, so that you can easily relate all results, is 2 V/division. To see the *total* voltage across gate-source and across drain-source, you need to use DC coupling for the scope inputs. If everything is working correctly, the variation of all these waveforms should be approximately sinusoidal; if it is not, *reduce the signal amplitude somewhat,* until the variation is approximately sinusoidal.

16. Sketch the above four waveforms. Make sure you understand the reason for their form. In particular, consider the waveforms for v_{GS} and v_{DS}; are these two related as you would expect from the plot of step 3?

17. Does the relation between the input and output waveforms justify the name *inverter* for this circuit?

18. Verify that the input signal appears amplified at the output. What is the amplification factor?

19. Is the amplification factor approximately equal to those obtained in steps 5 and 7? It should be. Why?

20. Vary the input signal amplitude slowly, and observe v_{DS} with the scope. How large can the sinusoidal variation in v_{DS} be made before it is visibly distorted? Why do you think it gets distorted? Try to answer this question by referring to the plot you obtained in step 3.

21. As we have seen, the bias values of v_{GS} and v_{DS} are the DC values around which the signal variations occur. We will now try other bias values and will study their effect. Reduce the input signal amplitude to zero. Adjust the value of V_{DC} until the transistor

is biased at $v_{DS} = 1$ V. Record the required value of v_{GS}. Then, increase the input amplitude somewhat, so that you can see a sinusoidal variation in v_{DS}. Determine the new amplification factor. Is it different from the one determined in step 18? Why? Answer this question by referring to the plot of step 3.

22. Increase the input signal amplitude, and determine how large the sinusoidal variation in v_{DS} can be made before significant distortion sets in. Is the result different from that in step 20? Why? To answer this question, again refer to the plot you obtained in step 3.

23. What bias value should one select for v_{DS} so that the maximum possible output signal amplitude, without significant distortion, is obtained from this circuit? Give the maximum output amplitude and the voltage gain for that bias point.

24. What is the range of signal frequencies that can pass through this amplifier? Take the lowest and highest usable frequencies to be the ones at which the voltage gain drops to $1/(\sqrt{2}) \approx 0.707$ of its value at 1 kHz. Audio amplifiers are often required to pass signals from 20 Hz to 20 kHz. Is this amplifier suitable for audio?

25. (Optional) If you have time, you can try to pass a music signal through your amplifier and feed the output to the power amplifier and speaker. The setup will be similar to that used in Experiment 5, except that here you would use a CD or cassette player as the signal source and your one-transistor amplifier rather than the op amp amplifier. You are on your own for this step.

NOTE: The circuits used in this experiment are kept simple to explain some basic principles. Although the method used above to bias the transistor is sometimes employed, this method turns out to be rather sensitive to temperature variations and device tolerances. You can learn how to design better biasing circuits in electronics classes.

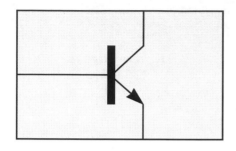

EXPERIMENT 13

BIPOLAR TRANSISTORS AND AMPLIFIERS

Objective

In this experiment you will study the *i-v* characteristics of an npn bipolar junction transistor. You will bias the transistor at an appropriate point for use as an amplifier and will try it as a microphone preamplifier. Some of the steps here parallel those in Experiment 12; if that experiment has been done, your instructor may ask you to skip these steps.

BACKGROUND

The bipolar junction transistor (BJT) is a three-terminal device often used in amplifiers and other electronic circuits. In this experiment we will work with a version of the device known as the *npn* transistor (the other version, which we will not consider here, is the pnp transistor). We will consider only the external *i-v* characteristics of the device; its physics and operation are studied in semiconductor device classes.

The symbol for a npn transistor is shown in Fig. 1. Assume that a positive voltage v_{CE} of at least several tenths of 1 V is applied between the collector and emitter. A collector

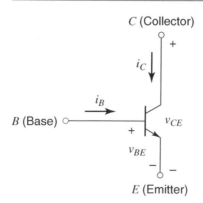

Fig. 1

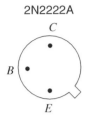

2N2222A

Bottom view

Bottom view

Fig. 2

current i_C can flow, depending on the voltage v_{BE} between the base and emitter. In other words, v_{BE} controls the value of i_C. The dependence of i_C on v_{BE} is *exponential*. Thus, the collector current can vary drastically as the base-emitter voltage is varied, and this makes possible the realization of high-gain amplifiers, as we will see in this experiment.

In addition, the presence of v_{BE} causes a small current i_B to flow into the base. This is usually a small fraction of the collector current i_C. The ratio i_C/i_B is called the current gain of the transistor, and it is often denoted by β_F. A typical value of this quantity is 100. It should be kept in mind, though, that this parameter is *not* reliable; it varies significantly with temperature, and can even be significantly different between two transistors of supposedly the same type.

In the first part of this experiment, we will view i_B as the controlling variable for i_C, to provide some familiarity with a common way of showing the transistor charac-teristics (and also to avoid repetition of Experiment 12, in which a voltage is used as the controlling variable for the MOS transistor). We should note, however, that when designing circuits it is more appropriate to view v_{BE} as the controlling variable; this view is more directly connected to the way transistors are (or should be) used in most circuits. We will switch to this latter point of view when using the transistor as an amplifier.

The BJT we will be using in this experiment is the 2N2222A. Lead assignments for two common packages for this device are shown in the bottom views of Fig. 2. In the package you are using, the lead assignment may be different; check with your instructor to make sure you know which lead is which in your transistor.

BIPOLAR TRANSISTOR *I-V* CHARACTERISTICS

1.   You will measure the transistor characteristics, using the setup in Fig. 3. The part labeled "Measuring setup" is a piece of instrumentation designed for making transis-tor measurements. The voltage of PS1 is used for varying the base current, i_B. The resistor is used for limiting this current (connecting PS1 directly across the base emit-ter would be risky since the base current depends exponentially on the base-emitter

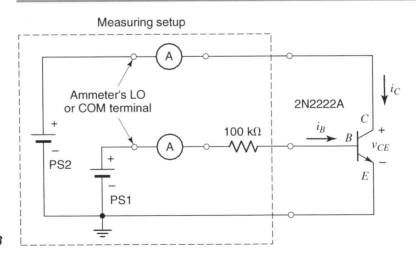

Fig. 3

voltage; thus a slight increase in the latter could cause a huge change in the base current). The circles labeled A are ammeters, used to monitor the base and collector currents. Hook up the circuit, *using connections as short as possible*. To avoid some problems associated with nonidealities of digital ammeters, connect them as shown in the figure. Note that when hooked up in this way, the ammeters will not show the base and collector currents but, rather, their opposite. Thus, you should change the sign of these readings to obtain the two currents.

2. Measure the collector current, i_C, versus the collector-emitter voltage, v_{CE}, from 0 to 10 V, for a fixed value of i_B, equal to 10 μA, using the following procedure:

- Set PS2 for a desired value of v_{CE}.

- If necessary, readjust PS1 so that $i_B = 10$ μA.

- Read the corresponding value of i_C.

By repeating this procedure for several v_{CE} values, you can form a table of i_C versus v_{CE}, which you can use to generate a plot later on. In the region where i_C is almost constant with respect to v_{CE}, you will need only a very few points. However, be sure you take a sufficient number of readings in the region where i_C is a strong function of v_{CE}; you will need these later to produce a smooth plot.

3. Repeat step 2 for i_B values of 20 μA, 30 μA, and 40 μA.

4. Using the data you have collected in the previous two steps, plot a family of curves for the collector current, i_C, versus the collector-emitter voltage, v_{CE}, from 0 to 10 V, with i_B as a parameter. Use a *single* set of axes for this plot. There should be a total of four curves on this plot (one for each of the four values of i_B). Label the plot properly. What is the approximate value of the current gain, $\beta_F = i_C/i_B$? We emphasize that this is not a value one can rely on. If the temperature changes, so will β_F.

A BJT-RESISTOR INVERTER

5. You will next use the circuit of Fig. 4, but *do not hook it up yet*. This circuit is an example of a setup that is *very* sensitive because, as you will see, the transistor can amplify the effect of voltage variations present between its base and emitter. Thus, if interference finds its way to the base or emitter, it will be amplified and can produce a large error, or puzzling readings, in the collector-emitter voltage. To minimize this possibility, follow these guidelines:[1]

- *Use connections as short as possible.*

- *Use only a single ground point*, as indicated in the figure. (If you use multiple ground points, the currents flowing from one such point to the next can cause minute voltage drops across the wires, which are not perfect short circuits. These minute

---

[1] If, after following these guidelines, you still encounter interference, consider the use of supply bypassing as described in Appendix D.

Fig. 4

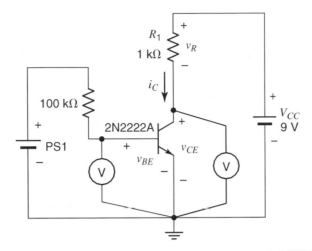

voltage drops can appear as variations in the base or emitter voltage; when amplified by the circuit, they may not be so minute anymore, and they can interfere with proper measurements.)

Following these guidelines, set up the circuit of Fig. 4, but do not turn on the power supplies yet.

6. By varying PS1, v_{BE} can be varied. When this voltage is raised sufficiently (e.g., to *roughly* 0.7 V for silicon transistors, such as the one we are using), the collector current i_C of the transistor will become significant for our purposes. This current will cause a voltage drop $v_R = R\, i_C$ across the load resistor R. Using KVL you can see that the collector-emitter voltage v_{CE} will be equal to $V_{CC} - v_R$ and will thus be less than V_{CC}. As v_{BE} is increased, i_C and thus v_R will also increase, and v_{CE} will decrease. Given that i_C is exponentially related to v_{BE}, a very slight increase of the latter can cause a drastic decrease in v_{CE}. Eventually, v_{CE} can become so small that it will limit further increases in i_C and thus in v_R. Based on this description, answer the following question without doing the experiment: What do you think the plot of v_{CE} versus v_{BE} would look like?

7. Turn on the power supplies. Take measurements and produce a plot of v_{CE} versus v_{BE} (the latter *should be varied by varying PS1*, and the corresponding values of v_{BE} and v_{CE} should be recorded). Be sure that you take enough measurements to adequately reproduce the steep part of the plot. Does the plot agree with your prediction in step 6?

8. In the plot produced in the previous step, consider the point at which $v_{CE} = 4$ V. What is the corresponding value of v_{BE}?

9. Your plot should be steep around the point defined in the previous step. If v_{BE} varies around this point, the corresponding variation of v_{CE} should be much larger. What is, approximately, the value of the slope $(\Delta v_{CE})/(\Delta v_{BE})$ around this point? Is it positive or negative? Why? The behavior you see is often said to be *inverting*. Why this name?

10. Set PS1 so that $v_{CE} = 4$ V. This will be the bias value of v_{CE}. Similarly, the corresponding value of v_{BE} is the bias value of that quantity. According to what you found in the previous step, if you can couple a small signal across base-emitter, so that it changes v_{BE} *around* its bias value, you will obtain an amplified version of this signal as a variation in v_{CE}. This is what you will do in the following step.

OBTAINING AC AMPLIFICATION WITH A BJT

11. To couple the small signal across base-emitter and to couple the resulting v_{CE} variation to an external device without disturbing the DC bias values set in step 10, use two coupling capacitors[2] as shown in Fig. 5. **If a polarity is indicated on the capacitors, make sure you connect them as indicated.** The load indicated as a resistor by a broken line is anything connected to the output, for example, a scope probe (the latter has a very large resistance value, so it will not appreciably affect the operation of the circuit). Measure the base-emitter and collector-emitter voltages again to make sure that they were not inadvertently changed.

12. Try the circuit as an amplifier by connecting the function generator with a 1 kHz, 5 mV peak signal at the input. Use the scope to study the waveforms of v_{IN}, v_{BE}, v_{CE}, and v_{OUT}, as well as how these are related to one another. To see the total voltage across base-emitter and across collector-emitter, you need to use DC coupling for the scope inputs; to see only the AC variation of these voltages, you need to use AC coupling instead. If everything is working correctly, the variation of all these waveforms should be approximately sinusoidal.

13. Sketch the above waveforms. Make sure you understand the reason for their form.

14. Does the relation between the input and output waveforms justify the name *inverter* for this circuit?

15. Verify that the input signal is amplified at the output. What is the amplification factor?

16. Is the amplification factor approximately equal to the slope obtained in part 9? It should be. Why?

17. Vary the input signal amplitude slowly, and observe v_{CE} with the scope. At what output amplitudes does the output waveform get visibly distorted? Why do you think it gets distorted? Try to answer this question by referring to the plot you obtained in step 7.

18. Try various other bias points while observing v_{CE}. Each time, turn down the input signal amplitude and adjust PS1 so that v_{CE} attains a desired DC (bias) value. Then,

---

[2]If you need more explanations about signal coupling through capacitors, you can study the section "Adding a signal to a DC bias voltage" in Experiment 12.

Fig. 5

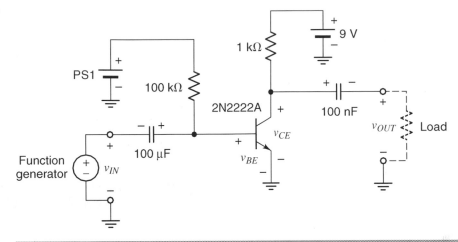

---

increase the input amplitude and determine how large the sinusoidal variation in v_{CE} can be before significant distortion sets in.

19. How should one select the bias point so that the maximum possible output amplitude, without significant distortion, is obtained from this circuit? Leave the circuit biased at such a point.

20. Replace the function generator by the microphone, and observe the input and output waveforms.

21. **(Optional)** If you still have time, you can try this transistor amplifier in lieu of the op-amp-based mike preamp you had designed earlier (see Experiment 5) in a complete sound system, including microphone, transistor preamplifier, volume control, power amplifier, speaker, and a power source for the power amplifier. How does this new preamplifier sound?

NOTE: The circuits in this experiment are kept as simple as possible in order to explain some basic principles. Although the method used above to bias the transistor is sometimes employed, the resulting circuit turns out to be sensitive to temperature variations and device tolerances. You can learn how to design better biasing circuits in electronics classes.

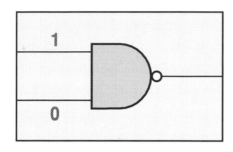

EXPERIMENT 14

DIGITAL LOGIC CIRCUITS; GATES AND LATCHES

Objective

In this experiment, you will become familiar with the characteristics of basic digital circuits. You will use logic gates to build a circuit that makes a logic decision by acting on input digital data. You will use logic gates to construct a memory element called a latch and a related circuit called an S-R latch.

Background

You will benefit from studying any basic introduction to logic gates and SR latches. Nevertheless, most of the background needed for this experiment will be provided along the way, as the individual steps are introduced.

LOGIC LEVEL INDICATOR

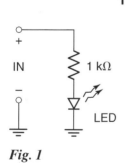

Fig. 1

1. In this experiment, you will need a logic level indicator, a circuit containing a light-emitting diode (or LED), which is on when the voltage provided to the indicator is sufficiently high and off if that voltage is sufficiently low. If you are using a breadboard designed for digital experiments, it will most likely include several logic level indicators, as well as "buffer circuits" (not visible to you) for each LED. The buffer circuits need only a very small current at their input, but they provide plenty of current to the LEDs, to turn them on brightly.

   If no preassembled logic level indicators are available or if your instructor suggests it, you can make your own indicator for this experiment as shown in Fig. 1. If the input voltage of this circuit is sufficiently large, enough current flows to make the LED light up. The resistor in series with the LED prevents the current from becoming excessive. Roughly how high an input voltage is needed for the LED to light up, assuming that it needs 2 mA of current and that the forward-bias voltage across it is 1.7 V? Set up the circuit and verify your prediction. Record two approximate input voltage values: a value below which the LED is fully off and a value above which the LED is strongly on. In our use of this level indicator, the exact input voltage value will be immaterial as long as the input stays below the first value (i.e., it is low) or above the second value (i.e., it is high). A low input corresponds to a logic 0; a high input corresponds to a logic 1.

We should note that some logic circuits may not be able to supply the current needed to turn on the LED in Fig. 1. The logic circuit family (the type of logic chips) we will be using in this experiment, however, can do so.

NAND GATES

2. Figure 2 shows a NAND gate with two inputs, A and B, and one output, C. Four such gates are contained in a 74LS00 chip, the pin assignment of which is shown in Fig. 3. *Connect this chip to a 5 V power supply*, as shown in the figure. Connect a voltmeter to the output of a NAND gate on the chip. In the following steps, you will be asked to apply an adjustable voltage at the input(s) of this gate and to observe and plot the behavior of the output voltage (at C in Fig. 2) as the input voltage is changed from 0 to 5 V. **Make sure that the input voltages do *not* exceed V_{CC} in *any* case.**

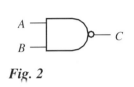

Fig. 2

3. Tie B to V_{CC} (+5 V). Connect a *separate* power supply voltage to A. Plot the voltage at C versus the voltage at A, with the latter varying from 0 to 5 V. Such a plot is known as a "DC voltage transfer characteristic."

4. On the plot you just obtained, input voltages below a certain critical level are said to be "low"; input voltages above a certain critical level are said to be "high". These critical levels are sometimes defined as the points where the plot has a slope of -1. Determine, approximately, these two levels from your plot.

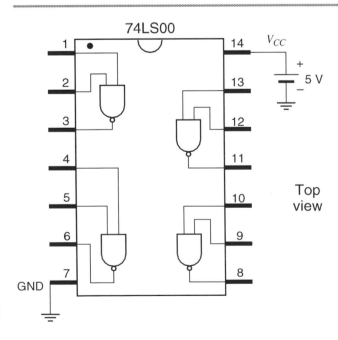

Fig. 3

Fig. 4

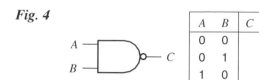

A	B	C
0	0	
0	1	
1	0	
1	1	

For steps 5 to 8 you do not need to produce plots; qualitative observation will do. Before starting each step, disconnect what you had connected to the gate's input in the previous step.

5. Tie B to ground (0 V) and observe the voltage at C as the voltage at A is varied from 0 to 5 V.

6. Tie A to V_{CC} and observe the voltage at C as the voltage at B is varied from 0 to 5 V.

7. Tie A to ground and observe the voltage at C as the voltage at B is varied from 0 to 5 V.

8. Tie both inputs A and B together and observe the voltage at C as the common input voltage is varied from 0 to +5 V.

9. As you saw in the previous parts, the output voltage varies continuously. However, in digital logic work we are only interested in whether the output is low or high. You will now retest the NAND gate in a digital fashion. Replace the voltmeter by your logic level indicator (provided on your breadboard, or from step 1). To provide input voltages corresponding to logic 0 or 1, connect the corresponding input pins to ground or V_{CC}, respectively. Try all four input combinations and observe the indication of the LED. Using a logic 0 to represent "low," and a logic 1 to represent "high," complete the *truth table* shown in Fig. 4 with logic 0s and 1s.

10. Does the table you just obtained justify the name NAND (i.e. NOT AND, the logic complement of the AND operation) for this gate? Why?

MAKING OTHER GATES OUT OF NAND GATES

11. The symbol and truth table for a NOT gate are shown in Fig. 5. How can you use a NAND gate as a NOT gate? Verify your answer experimentally using logic level indicators.

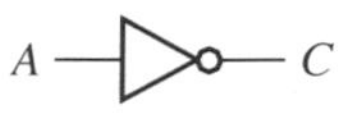

A ─▷o─ C

A	C
0	1
1	0

Fig. 5

Fig. 6

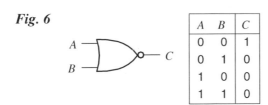

A	B	C
0	0	1
0	1	0
1	0	0
1	1	0

12.   The symbol and truth table for a NOR (i.e., NOT OR) gate are shown in Fig. 6. Explain this name. How can you make a NOR gate by using *only* NAND gates? *(Hint:* For two logic variables A and B, it can be shown that $\overline{A + B} = \overline{A} \cdot \overline{B}$, where $+$ stands for the logic OR operation, $\cdot$ stands for the logic AND operation, and a bar stands for the NOT operation.)

Verify your answer experimentally.

LOGIC DESIGN

13.   Design a logic circuit that warns the driver of a two-door car when any door is open and the driver tries to start the car. The circuit should provide a logic 1 in this case (this logic 1 can be used to activate an alarm or warning light). Assume a circuit installed at each door provides a logic 1 if that door is open and a logic 0 if that door is closed; also, a circuit connected to the starter provides a logic 1 when the starter is activated and a logic 0 when it is not. Use only NAND gates.

14.   Try your circuit. Use your logic level indicator as the warning light. Demonstrate your circuit to your lab instructor.

SIMPLE LATCH

15.   The circuit shown in Fig. 7, consisting of two NOT gates, is a common memory element called a *latch*. Verify theoretically that there are two states possible for this circuit: $Q = 1$, $\overline{Q} = 0$; and $Q = 0$, $\overline{Q} = 1$. The output $\overline{Q}$ is called the *complement* of Q.

16.   Construct this circuit, implementing the NOT gates by using NAND gates (see step 11). When you turn on the power supply, the circuit should go into one of the two states described above. Verify that you can change the state at will by momentarily shorting to V_{CC} the output that happens to be low, or by momentarily shorting to ground the output that happens to be high. Be sure you understand why this occurs.[1] Clearly, there are two distinct *states* that this circuit can memorize: $Q = 1$ and $Q = 0$. These two states can be maintained indefinitely if desired (as long as the power supply remains connected), thus keeping the information about the state the

---

[1]Shorting an output to ground or to the supply voltage is, in general, not a good idea, and can permanently damage some chips. However, it is acceptable to do so with the chips used in this experiment, for instructional purposes.

Fig. 7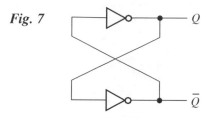

circuit was put in some time in the past. This circuit forms the basic principle of many static digital memories.

SR LATCH

17.   A better way to set the state of a latch is shown in Fig. 8. This is an $\overline{S}\ \overline{R}$ *latch,* implemented by using NAND gates. Verify theoretically the first three rows of the state table shown in Fig. 9, as follows. For the second and third rows, use the truth table for the NAND gate (Fig. 4) to arrive at the output values given; show that those values are the only ones consistent with the operation of the NAND gates. To verify the first row, assume that the two inputs of the latch were originally 0 and 1, or 1 and 0; show that if the 0 input is raised to 1, nothing will change at the outputs of the flip-flop; in other words, the latch will retain its previous state.

18.   You now need a convenient way for providing the input voltages to the circuit of Fig. 8. If you are using a breadboard designed for digital experiments, you can use the data switches on it to do what is required in step 19. These switches provide +5 V or 0 V (or other suitable values), depending on their position.

   Alternatively, you can use the circuit shown in Fig. 10. Use one such circuit for each input. The switch shown can be an actual switch (preferably a push-button one), or it can be just a wire connection that you can make or break. When the switch is closed, the output voltage of this circuit is zero. When the switch is open, the output voltage is equal to the power supply voltage. When the input of a logic gate is connected to this circuit, the current it will draw will cause a voltage drop across the resistor; however, this voltage drop will be small, and the voltage across the open switch will still be large enough to correspond to a logic 1.

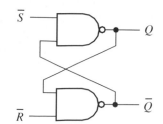

Fig. 8 $\overline{R}$

$\overline{S}$	$\overline{R}$	Q	$\overline{Q}$
1	1	No change	
1	0	0	1
0	1	1	0
0	0	Not used	

Fig. 9

Fig. 10

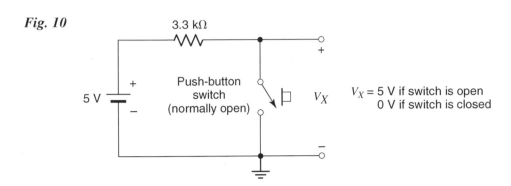

3.3 kΩ

Push-button switch (normally open)

5 V

V_X

V_X = 5 V if switch is open
0 V if switch is closed

+

−

19. Verify the first three rows of the function table of Fig. 9 experimentally. The two inputs to the latch should normally be kept high, and either one or the other should momentarily be switched to low to set or reset the latch.

It is clear that the latch can be set to the state $Q = 1$ or can be reset (i.e., moved to the state $Q = 0$), depending on which input is momentarily switched to zero. The "S" and "R" in the latch name stand for Set and Reset, respectively. Observe that once one of the two switches momentarily provides a logic 0, the latch is set or reset; after that point, pressing again the *same* switch has *no* effect on the state of the latch.

GATED SR LATCH

20. In the latch of Fig. 8, the outputs may change as soon as the inputs are changed. In some applications, this is not desirable. Instead, one may want to change the inputs to the desired values and *then* give a signal to the latch to change its output accordingly. Such a signal is provided by an "enable" input. To make such an operation possible, you need some additional circuitry in front of the latch, as shown in Fig. 11. You will be asked below to figure out what this circuitry is. The operation is to be as follows: First, desired values are applied at S and R. While the value at C is low, information about the inputs S and R is *not* allowed to pass to the latch; both latch inputs are 1 in

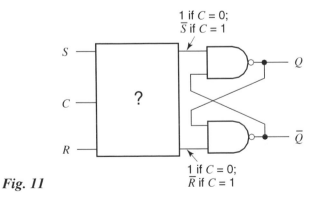

1 if $C = 0$;
$\overline{S}$ if $C = 1$

S

C

?

R

1 if $C = 0$;
$\overline{R}$ if $C = 1$

Q

$\overline{Q}$

Fig. 11

Fig. 12

S	R	C	Q
0	0	1	No change
0	1	1	0
1	0	1	1
1	1	1	Not used
×	×	0	No change

this case. As soon as C is made high, the *complements* of S and R are sent to the latch as indicated. This sets or resets the latch, according to what we have seen in step 19. The function table for this operation is shown in Fig. 12. In the first row of this table, the symbol × means a "don't care"; i.e., an input can be either 0 or 1. Figure out the circuitry inside the box in Fig. 11 to accomplish this. (*Hint:* The box contains only two NAND gates.) This circuit is called a "gated," or "clocked," *SR* latch.

21.   Build the complete circuit of Fig. 11. To apply a logic 1 to S or to R, tie that input to +5 V through a switch or wire. To apply a logic 0, tie the input instead to 0 V. If you have a digital breadboard with data switches, you can use one of them for the signal at C. Alternatively, you can use the circuit of Fig. 13; the switch should preferably be a push-button one. The switch is normally open, and the output of the circuit is zero; when the switch is pressed, the output becomes +5 V.

   Try the circuit. The inputs at S and R should not be allowed to change while C is high. Instead, with C low, set the values of S and R; then make C high, and observe the output. Next, set C low again, set new values for S and R, and make C high again; etc. The switch need only momentarily provide a logic 1, for the input data to determine the state of the latch. Verify all entries in the table shown in Fig. 12 except the one labeled "Not used."

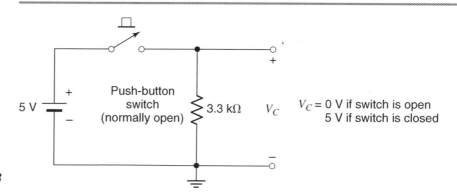

Fig. 13

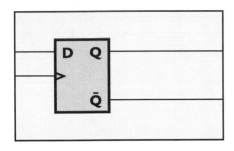

EXPERIMENT 15

D FLIP-FLOPS AND SHIFT REGISTERS

Objective
Here you will experiment with a widely used logic circuit: the edge-triggered D flip-flop. You will use this flip-flop to make a a shift register, a device into which you can enter data and then obtain them in serial form (i.e., one after another) or in parallel form (i.e., all at once).

Background
It is assumed that you have studied the basics of flip-flops and shift registers in your theory class. In case you do not feel very comfortable with them, some material is provided below to guide you along the way.

DATA INPUT, CLEAR, AND CLOCK SIGNALS

The devices we will work with are normally part of larger digital systems and receive logic signals from the rest of those systems. Here, however, to test such devices we will be producing our own logic test signals. If you are using a breadboard intended for digital experiments, these signals will normally be provided by preassembled circuits on your test board, as explained below. (If they are not, the required circuits can be assembled from scratch, as explained in Appendix E.) Be sure that the power to these circuits is on before testing their operation.

1. **DATA IN switch:** When we need voltages that correspond to a logic 1 or 0, and when the device to which these are fed is *not* sensitive to the details of the *transition* between these two levels, we can use the *data in* switch shown in Fig. 1(a). Depending on the position of the switch, a logic 1 or 0 appears at its output and *stays* there until the switch position is changed. An LED logic level indicator should be connected to the switch output; this indicator will normally be available on your digital experimentation breadboard. Verify the operation of the circuit by flipping the switch back and forth and observing the indication of the LED.

2. **CLEAR switch:** The *clear* switch in Fig. 1(b) provides a signal that is normally a logic 1. When the push-button switch is pushed, the output becomes a logic 0. As soon as the switch is released, the output goes back to its normal logic 1. *If the switch*

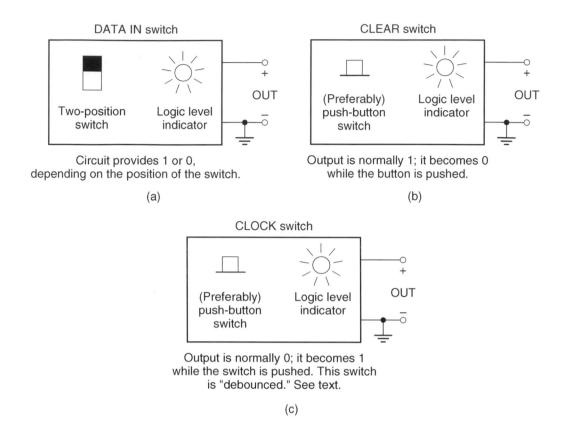

Fig. 1

on your breadboard is not a push-button one, maintain it at the logic 1 position; when you are asked to push (and release) the switch in the instructions below, you should move it to the logic 0 position and then back to the logic 1 position. The logic level provided by the switch should be indicated by an LED logic level indicator connected to it. Test the operation of this circuit.

3. **CLOCK switch:** The *clock* switch in Fig. 1(c) has an output that is normally 0 and becomes a "clean" 1 for as long as the switch is pressed.[1] *If your CLOCK switch is a two-position one, maintain it at the logic 0 position; when you are asked to push (and release) the clock button in the description that follows, move the switch to the logic 1 position and then back to the logic 0 position.* As before, an LED indicator should be connected to the output of the switch. Test the operation of this circuit.

---

[1] By "clean" we mean that the output is "debounced" by appropriate circuitry so that mechanical bouncing in the switch does not generate extraneous zeros and ones at the output. If you are interested, you can read the description of a debouncing circuit in Appendix E.

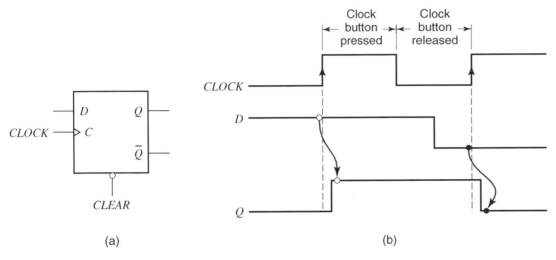

Fig. 2

EDGE-TRIGGERED D FLIP-FLOP

A *positive-edge-triggered D flip-flop* is shown in Fig. 2(a). Assume first that the *CLEAR* signal is high. Then the flip-flop has the following property: *Data at the input,* D, *is transferred to the output,* Q, *when the clock signal makes the transition from low to high.* That is, if the input D is steady at a logic 1 when the clock goes to high, the input logic 1 is transferred to the output Q; if, instead, the input is 0 when the clock goes to high, this 0 is transferred to the output. The data is held at the output until the next low-to-high clock transition. These properties are illustrated in Fig. 2(b). In this figure, we also indicate the fact that the new output appears after a very small delay (typically in the tens of ns). Since the input data are transferred to the output when the clock goes from low to high, this is called a positive-edge-triggered D flip-flop. For correct operation of this flip-flop, the data at D must be steady before the clock low-to-high transition takes place, at least for the minimum time specified by the flip-flop manufacturer (e.g., for 20 ns).

No data transfer takes place when the clock signal is steady or when it goes from high to low.

The output Q can be cleared by momentarily switching the *CLEAR* input to a logic 0. In all of the cases described, the second output, Q, is the complement of Q.

4. Hook up a D flip-flop as shown in Fig. 3, using one of the four D flip-flops inside a 74LS175 chip, the pin assignment of which is shown in Fig. 4. *Be sure you connect the ground and supply pins.* A logic level indicator should be used to provide a convenient indication of the flip-flop's output Q. If such an indicator is not available on your breadboard, make one as shown in Fig. 5.

5. Clear the output by pushing and then releasing the *CLEAR* button; Q should be 0 as a result of this operation. Now set D to 1. Push and then release the clock button. The 1

Fig. 3

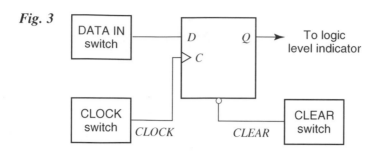

Fig. 4

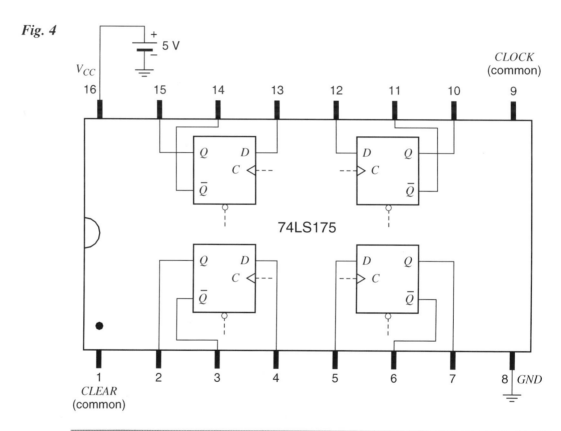

at D should be transferred to Q. When did this transfer occur? When the clock went high or when it went low? If you didn't notice, clear the output and repeat.

6. With a logic 1 at Q and the clock kept at 0, change the value of D to 0. Does the output change?

7. With Q at 1 and D at 0, push and then release the clock button. The 0 should be transferred to Q. At which clock transition did this occur? Low to high, or high to low? If you didn't notice, repeat the step.

8. With a logic 0 at Q and the clock kept low, change the value of D to 1. Does the output change?

Fig. 5

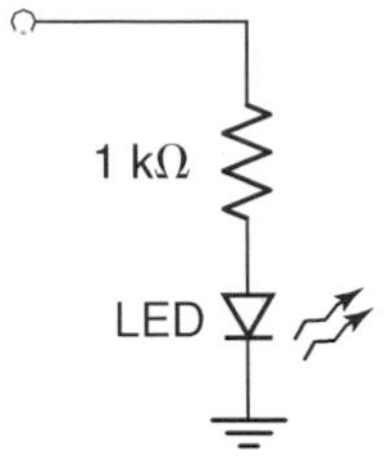

Fig. 6

D	C	Q
0	↑	0
1	↑	1

9.   Verify that the function table in Fig. 6 agrees with your observations above. An up-pointing arrow means a transition of the clock from low to high, and Q is the output value following that transition.

A SHIFT REGISTER

10.   Consider the four-stage shift register in Fig. 7, *but do not build it yet*. Assume that all outputs are cleared and then a 1 is placed at the input *IN*. Assume that the clock button is pressed and released *once*. What will the values of Q_0, Q_1, Q_2, and Q_3 become? [*Hint*: When the clock goes from low to high, each flip-flop transfers to its Q the value that existed at *its D just before* the clock transition; see Fig. 2(b).]

Now assume that after the above operation, the input *IN* is set to 0 and then the clock button is pressed once again. What will the new values of Q_0, Q_1, Q_2, and Q_3 become?

11.   Now build the circuit of Fig. 7. All four D flip-flops needed can be found inside the same 74LS175 chip, the pin assignment for which was given in Fig. 4. It will help if, before starting to build the circuit, you mark Fig. 7 with the appropriate pin numbers from Fig. 4. *Be sure to connect the ground and supply pins*. If no logic level indicators are provided on your breadboard, use four indicators of the type shown in Fig. 5.

12.   Clear the outputs of all flip-flops by pushing and then releasing the *CLEAR* button. Perform the operations described in step 10, and verify your predictions in that step. It should be clear that each time the clock goes from low to high, the 1 you had placed initially at the input is propagated one position to the right. Now press the *CLOCK* button again and again. What happens eventually? Why?

13.   Clear the register again. Place a 1 at the input and leave it there (i.e., do not change it to 0 after you hit the clock button). Push the clock button several times. What do you observe? Why does this occur?

14.   **Serial in, parallel out:** Clear the output. Set $IN = 1$; push and release the clock button. Keep $IN = 1$; push and release the clock again. Then set $IN = 0$; push and release the clock once again. Now the three consecutive values of the input 1, 1, and 0 must have

To logic level indicators

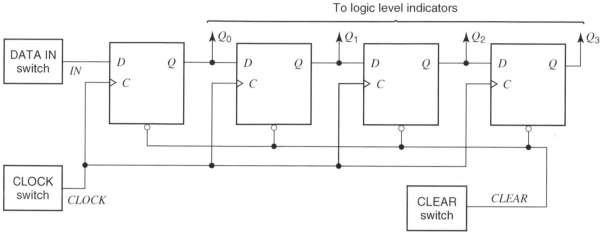

Fig. 7

been transferred to Q_2, Q_1, and Q_0, respectively. You entered the data serially; now they are available in parallel form, that is, simultaneously at the above three outputs. They can be picked up by digital logic for further processing.

15. **Serial in, serial out:** You have already entered the input values 1, 1, and 0 into the shift register. Now look only at the LED that indicates the state of Q_3 (cover the other LEDs). Push the clock button three times, each time observing the state of the output Q_3. You should see the data you had stored come out of the shift register serially, that is, one by one, in the same order they went in. Explain why this occurs.

16. Repeat the previous step for other input data sequences. Each time, start by clearing the outputs; then store the desired values in the shift register, and get them out one by one by pushing the clock button repeatedly. Try to anticipate what will occur before you push the clock button. Be sure you understand what is happening.

A CIRCULATING LIGHT

17. Can you make a circuit in which a single light circulates among four LEDs? We will think of these LEDs as arranged in a circle, although it may not be convenient to physically arrange them in this way. Only *one* LED should be on at any given time. Each time you push the clock button, the light should advance one position, thus going in circles every four clock hits. Try to think how you can make this circuit, and then go ahead and try it.

Instead of generating the clock by manually pushing a button, you can have a periodic clock waveform do the job. You will do so in the following steps.

18. If you have a generator with "TTL" output on your digital breadboard or function generator, you can use this output to produce a *CLOCK* waveform. Otherwise, set up your function generator to produce such a signal, as described in Appendix F. In either case, do *not* connect the generator to your circuit yet. To test the generator signal, observe the output of the generator with the scope, using the AUTO trigger mode. Set the period of the signal to 1 ms and the sweep rate on the scope to 1 ms/div. The output of the generator should consist of alternating high and low values, which are suitable for driving the circuits you are working with.

19. Lower the frequency of the signal to 1 Hz. *Disconnect the clock switch circuit from the input of the shift register of Fig. 7,* and connect instead your function generator output to it, as shown in Fig. 8. The LED at the function generator output should be lighting up 1 time per second.

20. If necessary, start your circuit as you have determined in step 17 (you may want to slow down the clock to conveniently do this step). The light should now be circulating. Try other clock periods to make the light go around faster. Demonstrate the circuit to your instructor.

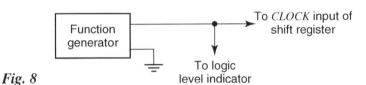

Fig. 8

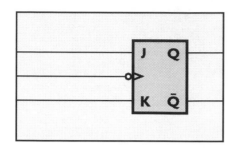

EXPERIMENT 16
JK FLIP-FLOPS AND RIPPLE COUNTERS

Objective   In this experiment you will work with JK flip-flops, and you will use them to build a ripple counter, a device that counts the number of events at its input.

Background   A simple introduction to JK flip-flops and ripple counters in your theory class should be sufficient for you to do this experiment. Some material will be provided below to refresh your memory with the basics of these circuits.

CLEAR AND CLOCK SIGNALS

In this experiment we will be using circuits that include switches to produce *CLEAR* and *CLOCK* signals. Read the description of these circuits in the first part of Experiment 15. If you are using a digital experiment breadboard, these circuits will normally be available on it (if they are not, the required circuits can be assembled from scratch, as explained in Appendix E). Turn on these circuits and test their operation.

The instructions below have been written assuming that your *CLEAR* and *CLOCK* switches are push-button types. If, instead, your switches are two-position ones, maintain them in their normal position (see Experiment 15, Fig. 1) and, when you are asked to "push" or "hit" them, switch them momentarily to their other position and then back to their original one.

LOGIC LEVEL INDICATORS

Logic level indicators should be available on a digital breadboard. Otherwise, you can make these indicators as shown in Experiment 15, Fig. 5.

JK FLIP-FLOP

A negative-edge-triggered JK flip-flop is shown in Fig. 1. Assume first that the *CLEAR* and *PRESET* inputs are high. Then the output can change at the *high-to-low*

Fig. 1

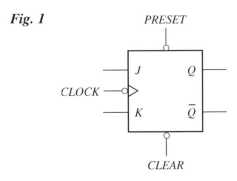

Fig. 2

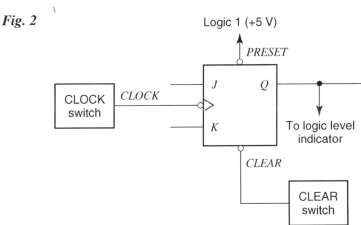

transitions of the *CLOCK*. At such transitions, if $J = 1$ and $K = 0$, the flip-flop is set; that is, Q goes to 1. If, instead, $J = 0$ and $K - 1$, the flip-flop is reset; that is, Q goes to 0. If both $J = K = 0$, the output maintains its previous value; finally, if $J = K = 1$, the output assumes the *complement* of its previous value.

Both the *CLEAR* and *PRESET* inputs are normally kept high. The flip-flop output can be cleared to 0 by momentarily making *CLEAR* low. (It can also be preset to 1 by momentarily making *PRESET* low, but we will not use this feature.) In all cases, the second output, $\overline{Q}$, is the complement of Q.

1. Hook up the circuit of Fig. 2. Use either one of the two JK flip-flops inside a 74LS112 chip, the pinout of which is shown in Fig. 3. *Be sure you connect the ground and supply pins.* The *PRESET* input should be permanently connected to a logic 1 (the +5 V supply voltage), as indicated in Fig. 2.

2. Verify the description of the JK flip-flop given above. Each time clear the output by momentarily pushing the *CLEAR* button; set the values of J and K by connecting them to the supply voltage or to ground as appropriate; then hit the clock button. Verify that the output changes at the *high-to-low* transition of the clock.

3. Verify that the function table of Fig. 4 agrees with your observations. A down-pointing arrow means a transition of the clock from high to low, with Q_{n-1} and Q_n being the output values before and after the transition, respectively.

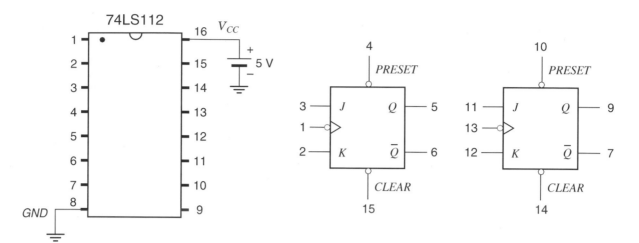

Fig. 3

Fig. 4

J	K	C	Q_n
0	0	↓	Q_{n-1}
0	1	↓	0
1	0	↓	1
1	1	↓	$\overline{Q}_{n-1}$

4. Tie both *J* and *K* to the positive power supply. Hit the *CLOCK* button repeatedly. The output should toggle, that is, alternate between 0 and 1 each time the clock signal goes from high to low.

A RIPPLE COUNTER

5. Consider the ripple counter shown in Fig. 5, *but do not build it yet*. Assume all outputs are cleared. Assume that the clock button is hit repeatedly, resulting in the clock waveform shown at the top of Fig. 6. Complete the waveforms for Q_0, Q_1, Q_2 and Q_3. (*Hint*: According to what you found above, *each* flip-flop output will toggle each time the clock input of *that* flip-flop goes *from high to low*.)[1]

6. Now build the circuit of Fig. 5. Two JK flip-flops can be found in one 74LS112 chip, so you will need two such chips. The pin assignment has been given in Fig. 3. You can mark the pin numbers directly on Fig. 5 to make connecting easier. *Be sure you connect*

---

[1]In logic design classes you will learn that the practice of driving the clock input of a flip-flop from the Q output of a previous flip-flop (as opposed to driving it from system clock) can present some problems. This approach, however, is adequate for what we want to demonstrate here.

To logic level indicators

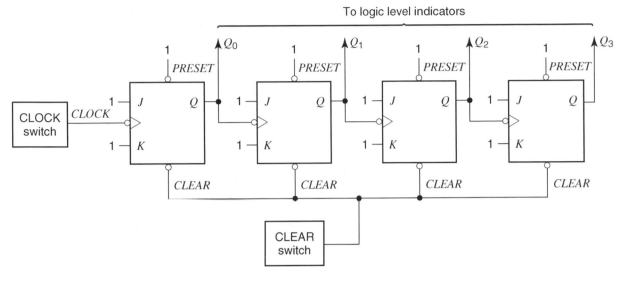

Fig. 5

Fig. 6

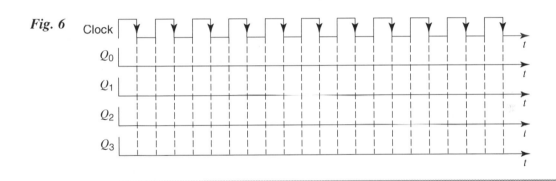

the ground and supply pins. The *PRESET* inputs should be permanently connected to a logic 1, as shown in Fig. 5.

7. Clear the outputs of all flip-flops by pressing the clear button. Push the clock button repeatedly, each time observing the outputs of the four flip-flops. Verify your predictions in step 5.

8. Clear the outputs. Consider the combined outputs $Q_3 Q_2 Q_1 Q_0$ as a binary number, with Q_3 the most significant bit and Q_0 the least significant bit.[2] Verify that this number represents the number of times the clock button is hit (or, more precisely, the number of times the clock signal goes from high to low) after the outputs have been cleared. For example, if you hit the clock five times, the binary number $Q_3 Q_2 Q_1 Q_0$ should be 0101. Explain why this happens, with the help of your diagram in Fig. 6.

---

[2]The decimal value of this number will be $Q_3 \times 2^3 + Q_2 \times 2^2 + Q_1 \times 2^1 + Q_0 \times 2^0$, where any of the Q_i's can be 0 or 1.

9. Keep hitting the clock button until the above binary number is 1111. What happens if you hit the clock one more time? Why?

USING A PERIODIC CLOCK

10. If you have a TTL output on your breadboard or function generator, you can use this output to produce a *CLOCK* waveform. Otherwise, set up your function generator to produce such a signal, as described in Appendix F. In either case, do *not* connect the generator to your circuit yet. To test the generator signal, observe the output of the generator with the scope, using the AUTO trigger mode. Set the period of the signal to 1 ms and the sweep rate on the scope to 1 ms/div. The output of the generator should consist of alternating high and low values, which are suitable for driving the circuits you are working with.

11. Lower the frequency of the signal to 1 Hz. *Disconnect the clock switch from the input of the ripple counter of Fig. 5,* and connect instead your function generator output, as shown in Fig. 7. The LED indicator at the output of the function generator should be lighting up one time per second.

12. Clear the ripple counter by momentarily pressing the CLEAR button. The counter should now be counting the number of clock pulses, as before. Verify that the binary word $Q_3 Q_2 Q_1 Q_0$ assumes the values 0000, 0001, 0010, and so on, up to 1111 and then back to 0000 again.

OBSERVING THE WAVEFORMS

13. Disconnect the LEDs connected to the inputs and outputs.[3] Increase the function generator frequency to 1 kHz. Observe the input signal on channel 1 of the scope, using a × 10 probe. Adjust the scope sensitivity to obtain a waveform of a reasonable size on the screen. Use a sweep rate of 1 ms/division. Trigger from channel 1. Use the AUTO trigger, and adjust the trigger controls for a stable display.

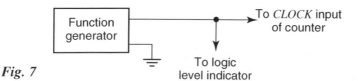

Fig. 7

---

[3]The clock frequency in the rest of this experiment will be high (1 kHz), and thus the LEDs cannot serve as optical indicators of ones and zeros. If you left them connected, the rapid succession of ones and zeros would result in a partially bright indication, which can be confusing.

14. Use the second channel of the scope and another $\times 10$ probe to observe the waveform at Q_0. Using the vertical position control of the scope, place this waveform below the input waveform. Verify that the two waveforms are related as you have drawn in Fig. 6.

15. Move the second probe to Q_1, and again observe the relation to the input waveform. Verify that it is as you have drawn in Fig. 6. Repeat for the waveforms at Q_2 and Q_3. At each of these outputs, the circuit is said to accomplish "frequency division." Why? What is the division factor for each output?

SIGNALING A PARTICULAR COUNT

16. Design a circuit that will make the LED flash every time a count of 14 is reached, that is, when the binary word $Q_3 \, Q_2 \, Q_1 \, Q_0$ reaches the value 1110. The LED should flash once and then turn off again as other binary values are reached. Build this circuit and demonstrate it to your instructor.

APPENDIX A
COMPONENT VALUE CODES

RESISTORS

A common way of labeling resistors is shown in Fig. 1 (this coding applies to resistors with a tolerance of 2% or higher; see below). There are four color bands. With the resistor positioned as shown, and going from left to right, the meaning of the first three color bands is indicated in the figure. Each color corresponds to a number, as shown in Table 1.

TABLE 1

Color	Number
Black	0
Brown	1
Red	2
Orange	3
Yellow	4
Green	5
Blue	6
Violet	7
Gray	8
White	9

Note in Fig. 1 that the interpretation of the third color band is *different* from that of the first two. As an example, assume that you positioned a resistor as shown and see, from left to right: yellow, violet, and orange. This means that the first digit is 4, and the second is 7, and there should be three zeros following these two digits. Thus the resistance value is 47000 Ω.

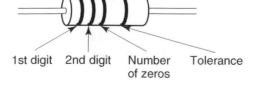

1st digit   2nd digit   Number   Tolerance
of zeros

Fig. 1

The fourth color band indicates the resistance tolerance. For this band, red means 2%, gold means 5%, and silver means 10%. No color means 20%. For example, if in the above resistor the fourth band is silver, the resistance value is 47000 ± 10%; that is, the resistor value is guaranteed to be somewhere between 42300 Ω and 51700 Ω.

Resistors of a specific tolerance come in standard values, for example, the 47000 Ω value above. The next higher available value in a 10% tolerance resistor series is 56000 Ω, which means that the actual value will be somewhere between 50400 Ω and 61600 Ω. As you can see, the range of values for each of the two resistors slightly overlaps. The standard values in a 10% tolerance resistor series are chosen so that this more or less occurs, which is why you encounter values that at first sight seem arbitrary. In a 10% tolerance resistor series, starting with a resistor of 1 Ω, the next values available are 1.2, 1.5, 1.8, 2.2, 2.7, 3.3, 3.9, 4.7, 5.6, 6.8, 8.2, 10, 12, 15, and so on. In a 5% tolerance series, there will be more values in between.

There are other, more detailed color-coding schemes for high-quality resistors. In this lab, we will not need to use them.

CAPACITORS

Unfortunately, common schemes for denoting capacitor values can be very confusing. Values can be written explicitly on capacitors (e.g., 50 μF) or more cryptically, 0.05 K or 102 M. Beware of the last letter—it is not part of the capacitance value, but rather indicates the tolerance. To read a capacitor value, *disregard* the last letter and interpret the remaining digits. These, to make our life difficult, give the value in either pF (10^{-12} F) or μF (10^{-6} F) but do not tell us which of the two it is. However, experience will tell you that 0.05 pF is too small a value to be available on a practical capacitor. Thus, the 0.05 K capacitor is a 0.05 μF one. In the case if the 102 M capacitor, the third digit represents the number of zeros, just as does the third color band in resistors. Thus, 102 means 10 followed by two zeros, or 1000. Unless the capacitor is physically huge, this cannot mean something as large as 1000 μF; hence, the value of the capacitor is likely to be 1000 pF = 1 nF (10^{-9} F). Unfortunately, one also sees capacitors with "1000" written on them, taken to mean "100 with 0 zeros," i.e., 100 pF! Other ways of denoting capacitance values also exist, including color codes.[1]

Other markings on a capacitor may be a code for the capacitor type; the maximum voltage it can handle; and, in the case of electrolytic capacitors, its polarity, which must be very carefully observed—otherwise the capacitor can explode.

---

[1]S. Wolf and R. F. M. Smith, *Student Reference Manual for Electronic Instrumentation Laboratories,* Prentice-Hall, Upper Saddle River, NJ, 1990.

APPENDIX B

OSCILLOSCOPE PROBE CALIBRATION

The $\times 10$ oscilloscope probe should be adjusted when viewing varying voltages so that the probe's own transient response does not affect the measurement. This is conveniently done by feeding a square wave to the probe (usually provided at an appropriate terminal on the scope's front panel) and adjusting the probe so that the waveform on the display is, indeed, square. Depending on the type of probe you are using, this adjustment may have to be done by rotating a little screw in it or by rotating part of the body of the probe. Your scope manual or your instructor can help you do this calibration properly. A properly calibrated probe will result in the waveform shown in Fig. 1(b), and not in those shown in Figs. 1(a) or 1(c).

If you are interested in knowing what exactly it is you are adjusting, read on. A $\times 10$ probe contains a resistor R_{probe} nine times larger than the input resistance of the scope R_{scope}, forming a voltage divider with the latter and feeding to the scope 1/10 of the signal. However, the scope input and the cable have unavoidable parasitic capacitance, C_{par}, to ground; in conjunction with these resistances, this creates a time constant that slows down the response of the probe, much like the slow charging effects studied in Experiment 6. To cancel this effect, an extra capacitor C_{probe} is used inside the probe body in parallel with the resistance of the probe, as shown in Fig. 2. This is the capacitor you are adjusting. (The strike through the capacitor symbol means that

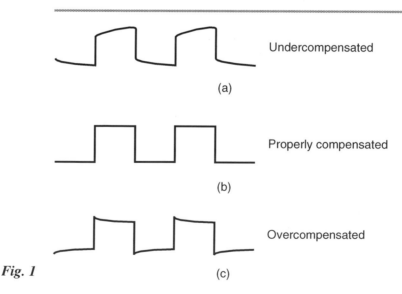

(a)   Undercompensated

(b)   Properly compensated

(c)   Overcompensated

Fig. 1

Fig. 2

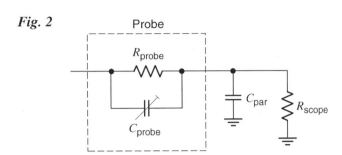

this component is adjustable.) It can be shown that when this capacitor is adjusted so that $R_{probe}C_{probe} - R_{scope}C_{par}$, the speed problems just mentioned are eliminated and a pure voltage divider action is possible (if you know about impedances, it is not difficult to prove this). A square wave fed to the probe, then, will appear as a square wave of one-tenth the size, without any exponential charging or discharging parts.

APPENDIX C

DECIBELS

Voltages and currents can be amplified or attenuated. For example, an amplifier might amplify a voltage by a factor of 1000, and a voltage divider might divide (attenuate) a voltage by a similar factor. The values encountered for such factors in circuits can vary over orders of magnitude. It is thus convenient to use a logarithmic measure for them. Let X be a voltage or current amplification or attenuation factor. A widely used logarithmic measure is

$$X_{dB} = 20\log X$$

The values of X_{dB} are expressed in units called decibels, or dB. Thus, for example, when $X = 0.1$, $X_{dB} = 20\log 0.1 = -20$ dB. Table 1 gives X_{dB} for a variety of X values. The values of ± 6 and ± 3 dB in the table are approximate; more accurate values are ± 6.02 dB and ± 3.01 dB. The quantity X_{dB} is sometimes referred to as "X in dB."

TABLE 1: VOLTAGE OR CURRENT FACTORS IN dB

X	X_{dB}
0.001	−60 dB
0.01	−40 dB
0.1	−20 dB
0.5	−6 dB
$1/\sqrt{2} \simeq 0.707$	−3 dB
1	0 dB
$\sqrt{2} \simeq 1.414$	+3 dB
2	6 dB
10	20 dB
100	40 dB
1000	60 dB

APPENDIX D

SUPPLY BYPASSING

Sometimes a circuit may not work properly because it receives interference from elsewhere (a radio or TV station, various types of lighting and equipment operated nearby, etc.). Such interference can be picked up by long wires connected to your circuit (e.g., those carrying power to your setup), which can act as antennas. In some cases, even the output of a power supply can contain interference.

To reduce interference, you may have to use *bypass* capacitors on your breadboard. A bypass capacitor is connected across a power supply voltage, at the point where this voltage enters your board or even closer to the circuit you are trying to protect. High-frequency interference signals then can go through the capacitor rather than through your circuit. Use a large capacitor (e.g., 10 to 100 μF; **be careful with its polarity as it can explode if connected the wrong way**). Such a capacitor can have a large internal parasitic inductance. If this interferes with proper bypassing at high frequencies, use in parallel with it a smaller one (e.g., 0.05 μF), of a type known to have low inductance (e.g., the ceramic type). This second capacitor helps bypass interference at very high frequencies. In professional systems, the inclusion of supply bypass capacitors is very common. Small resistors (e.g., 10–100 ohm) are sometimes included in series with the supply line, as shown in Fig. 1. Sometimes separate bypass circuits are used for separate circuits run from the same power supply. Each RC combination forms a low-pass filter (Experiment 9), which allows the DC voltage to pass but rejects high-frequency interference. This use of separate bypass circuits also helps to prevent interference generated by one circuit from being coupled to other circuits.

The element values above may have to be modified, depending on the amount and type of interference, the DC current drawn by the circuits, and so on.

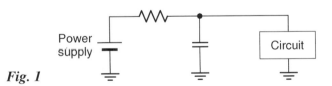

Fig. 1

APPENDIX E

DATA INPUT, CLEAR, AND CLOCK SWITCHES

To do Experiments 15 and 16, we need certain logic signals produced by switches and associated circuitry. Such circuits are normally provided on digital circuit training boards. If this is not the case with the board used in your lab, the circuits can be assembled from scratch, as explained below (Experiment 15 needs all three circuits; Experiment 16 needs only the clear and clock switch circuits). All three circuits can share the same power supply. Because of the length of Experiments 15 and 16, it would be better for these circuits to be preassembled before the start of the lab session.

1. **DATA IN switch circuit:** When we need voltages corresponding to a logic 1 or 0 and when the device to which these are fed is *not* sensitive to the *transition* between the two, we can use the simple circuit of Fig. 1. The switch has two stable positions; that is, it is *not* of the push-button variety. Depending on the position of the switch, a logic 1 or 0 appears at its output, and *stays* there until the switch position is changed. The state of the output is conveniently indicated by the LED, which should preferably be mounted close to the switch for easy association with the latter during the experiment. Connect this circuit and test its operation by flipping the switch and observing the indication of this LED.

2. **CLEAR switch circuit:** The circuit of Fig. 2 provides a signal that is normally high (logic 1). When the switch (preferably a push-button type) is open, the voltage developed across the output is large enough to qualify as a logic 1. When the push-button switch is pushed, a short is applied across the output of the circuit, and the output voltage drops to zero (logic 0). As soon as the switch is released, the output goes back to its normal logic 1. Build this circuit and test its operation.

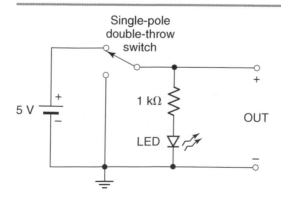

Fig. 1

Fig. 2

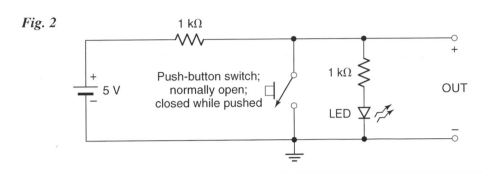

Depending on where they are used, the circuits of Figs. 1 and 2 have a potential problem. When a switch is thrown or pressed to make a contact, the switch can mechanically bounce several times over a short period (e.g., 1 millisecond), thus quickly closing and opening, before it finally makes a good contact. The output voltage of a circuit with such a switch can therefore quickly alternate between low and high a few times before it settles. If that output were to be fed directly to certain logic circuits, the circuits could interpret the resulting downs and ups of the voltage as alternating zeros and ones and could act on such false input "data". We will be using these circuits only when the circuits fed by them are not sensitive to such problems.

3. **DEBOUNCED CLOCK switch circuit:** Voltages generated by throwing switches that can bounce can be "cleaned," or "debounced," by using an SR latch, as shown in Fig. 3. This circuit is similar to one used in Experiment 14, and can be understood through that material. The switch is normally at A, making the voltage there 0 and the voltage at B high. When the switch is pressed, it momentarily touches B and then bounces. The first of the down transitions in the voltage at B sets the flip-flop, and further ups and downs in that output do not change the flip-flop's state. Similarly, when the switch is released, it touches A and then bounces; the first of the down transitions of the voltage at A resets the flip-flop. Thus, the output voltage of the circuit is a clean 1 or 0 and is oblivious to switch bouncing. This voltage can be used to drive other digital circuits. To conclude, then, the output of this circuit is normally 0 and becomes a "clean" 1 for as long as the switch is pressed.

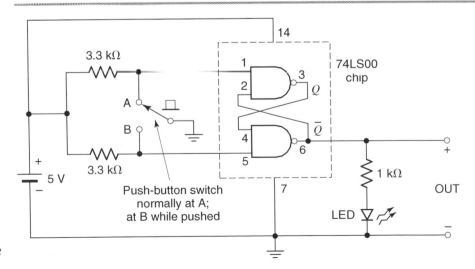

Fig. 3

APPENDIX F

PRODUCING A CLOCK SIGNAL BY USING A FUNCTION GENERATOR

The last few parts of Experiments 15 and 16 require a clock signal, which is basically a square waveform as shown in Fig.1. The lows of the waveform should be at 0 or at a small positive value. The highs should be at +5 V or at a somewhat smaller value. If your testing setup does not make such a "TTL" waveform available, you will need to produce one by using the function generator, as follows.

Turn on your function generator and observe its output on channel 1 of the scope. Set the scope's input coupling to DC; trigger from channel 1, and set the trigger mode to AUTO. Use a sweep rate of 1 ms/div. Set the generator to produce a square wave with a period of 1 ms and a peak-to-peak amplitude of 5 V. Adjust the trigger controls on the scope for a stable display.

Locate the offset control on your generator and turn it on. As you vary this control, the output waveform should move up and down on the scope screen. Adjust this control until the lows of the waveform are at 0. To make sure this is the case, establish the 0 level on the screen by using the ground input mode on the scope; then return the input coupling to DC, and be sure that the lows of your waveform are at this level. If necessary, readjust the amplitude and, possibly, the offset again until the output of the generator varies between two levels: 0 and 5 V. The signal should *not* vary outside the range 0 to 5 V, as the chips (to which you will connect this signal) can be damaged. However, lows slightly *above* 0 V and highs somewhat *below* 5 V are acceptable.

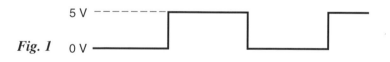

Fig. 1

APPENDIX G
EQUIPMENT AND PARTS LIST

> More details, as well as suggestions for distributors and for selecting equipment models, are given in the Instructor's Manual and/or the book's Web site, **www.wiley.com/college/tsividis**

Prototyping equipment

- Proto or other board, preferably mounted on a larger board with connectors. Several versions are available as "trainer" boards.
- A front panel for mounting a potentiometer and a variable capacitor.
- Wire kit.
- Long-nosed pliers.
- IC extractor.

Measuring equipment

- Two digital multimeters.
- 20 MHz Oscilloscope and oscilloscope probes ($\times 1$, $\times 10$).
- Two 1 Hz–2MHz function generators (one with modulation capability). A simple function generator may be available on a trainer board.
- Two dual, regulated variable 0–15 V power supplies. A simple power supply may be available on the trainer board.

Other equipment

- 1 W power amplifier with a voltage gain of 1 (easily made or available in kit form; see the Instructor's Manual and/or Web site).
- Loudspeaker with magnetically shielded enclosure and removable front panel.
- Dynamic microphone.
- CD player (preferably with volume control).

Cables, connectors, and adaptors

- A variety of connecting cables with appropriate end connectors, depending on the equipment used.
- Connector-to-connector adaptors.

Resistors

- ½ W, 5% or 10% tolerance; standard values from 10 Ω to 10 MΩ.

Capacitors

- 10% tolerance, 35 V; standard values from 100 pF to 100 μF.

Inductors	■ Radio frequency transformer wound around ferrite core, for AM radio inputs (several hundred μH). (E.g., ELENCO 484004.)
Variable capacitor	■ The variable capacitor should tune the AM band in conjunction with the above inductor (several hundred pF). (E.g., ELENCO 211677.)
Potentiometers	■ Two 10 kΩ linear potentiometers.
Semiconductor devices	■ Silicon power diode (1N4007 or equivalent). ■ Germanium small-signal diode (1N34A or equivalent). ■ Thermistor – several kΩ (type noncritical). ■ Photoresistor (type noncritical). ■ LED (low current; type noncritical). ■ MOS transistor (in CD4007 complementary pair/inverter chip). ■ Bipolar transistor (2N2222A or equivalent).
Integrated circuits	■ 741 op amp. ■ CD4007 complementary MOS pair/inverter. ■ 74LS00 Quad NAND gate. ■ CD4066 Quad bilateral analog switch. ■ 74LS175 Quad D flip-flop. ■ 74LS112 Dual JK negative edge-triggered flip-flop.